Methods in Microbiology
Volume 43

Recent titles in the series

Volume 26 *Yeast Gene Analysis*
AJP Brown and MF Tuite

Volume 27 *Bacterial Pathogenesis*
P Williams, J Ketley and GPC Salmond

Volume 28 *Automation*
AG Craig and JD Hoheisel

Volume 29 *Genetic Methods for Diverse Prokaryotes*
MCM Smith and RE Sockett

Volume 30 *Marine Microbiology*
JH Paul

Volume 31 *Molecular Cellular Microbiology*
P Sansonetti and A Zychlinsky

Volume 32 *Immunology of Infection, 2nd edition*
SHE Kaufmann and D Kabelitz

Volume 33 *Functional Microbial Genomics*
B Wren and N Dorrell

Volume 34 *Microbial Imaging*
T Savidge and C Pothoulakis

Volume 35 *Extremophiles*
FA Rainey and A Oren

Volume 36 *Yeast Gene Analysis, 2nd edition*
I Stansfield and MJR Stark

Volume 37 *Immunology of Infection*
D Kabelitz and SHE Kaufmann

Volume 38 *Taxonomy of Prokaryotes*
Fred Rainey and Aharon Oren

Volume 39 *Systems Biology of Bacteria*
Colin Harwood and Anil Wipat

Volume 40 *Microbial Synthetic Biology*
Colin Harwood and Anil Wipat

Volume 41 *New Approaches to Prokaryotic Systematics*
Michael Goodfellow, Iain Sutcliffe, and Jongsik Chun

Volume 42 *Current and Emerging Technologies for the Diagnosis of Microbial Infections*
Andrew Sails and Yi-Wei Tang

Methods in Microbiology
Volume 43

Imaging Bacterial Molecules, Structures and Cells

Edited by

Colin Harwood
*The Centre for Bacterial Cell Biology
Newcastle University
Newcastle upon Tyne
United Kingdom*

Grant J. Jensen
*California Institute of Technology
Howard Hughes Medical Institute
Pasadena, CA
United States*

AMSTERDAM • BOSTON • HEIDELBERG • LONDON
NEW YORK • OXFORD • PARIS • SAN DIEGO
SAN FRANCISCO • SINGAPORE • SYDNEY • TOKYO
Academic Press is an imprint of Elsevier

Academic Press is an imprint of Elsevier
125 London Wall, London, EC2Y 5AS, United Kingdom
The Boulevard, Langford Lane, Kidlington, Oxford OX5 1GB, United Kingdom
525 B Street, Suite 1800, San Diego, CA 92101-4495, United States
50 Hampshire Street, 5th Floor, Cambridge, MA 02139, United States

First edition 2016

© 2016 Elsevier Ltd. All rights reserved.

No part of this publication may be reproduced or transmitted in any form or by any means, electronic or mechanical, including photocopying, recording, or any information storage and retrieval system, without permission in writing from the publisher. Details on how to seek permission, further information about the Publisher's permissions policies and our arrangements with organizations such as the Copyright Clearance Center and the Copyright Licensing Agency, can be found at our website: www.elsevier.com/permissions.

This book and the individual contributions contained in it are protected under copyright by the Publisher (other than as may be noted herein).

Notices
Knowledge and best practice in this field are constantly changing. As new research and experience broaden our understanding, changes in research methods, professional practices, or medical treatment may become necessary.

Practitioners and researchers must always rely on their own experience and knowledge in evaluating and using any information, methods, compounds, or experiments described herein. In using such information or methods they should be mindful of their own safety and the safety of others, including parties for whom they have a professional responsibility.

To the fullest extent of the law, neither the Publisher nor the authors, contributors, or editors, assume any liability for any injury and/or damage to persons or property as a matter of products liability, negligence or otherwise, or from any use or operation of any methods, products, instructions, or ideas contained in the material herein.

ISBN: 978-0-12-809392-4
ISSN: 0580-9517 (Series)

For information on all Academic Press publications visit our website at https://www.elsevier.com/

Publisher: Zoe Kruze
Acquisition Editor: Alex White
Editorial Project Manager: Hannah Colford
Production Project Manager: Vignesh Tamil
Cover Designer: Victoria Pearson

Typeset by SPi Global, India

Contents

Contributors ix
Preface xi

SECTION 1 FLUORESCENT MICROSCOPY

CHAPTER 1 Methods for Visualization of Peptidoglycan Biosynthesis 3
Y.-P. Hsu, X. Meng, M.S. VanNieuwenhze

1. Peptidoglycan Structure and Biosynthesis 4
   1.1. Peptidoglycan Structure and Configuration 4
   1.2. Enzymatic Pathway of Peptidoglycan Biosynthesis 6
   1.3. Structural Variations in Peptidoglycan Structure 7
2. Strategies for Peptidoglycan Imaging: Electron Microscopy 9
   2.1. Revealing Bacterial Morphology 9
   2.2. Elucidating PG Structure via EM 11
   2.3. Cryo-TEM: Studying Cell Wall Organization
        in Natural State 13
   2.4. Three-Dimensional EM Technique: Cryoelectron
        Tomography 14
   2.5. More Tools: AFM in Bacterial Cell Wall Studies 16
3. Strategies of Peptidoglycan Imaging: Optical
   Microscopy 18
   3.1. Optical Microscopy Operation and Application 18
   3.2. Fluorescent Molecular Probes for PG-Specific
        Labelling 21
   3.3. Development of Metabolic Probes for PG-Specific
        Labelling: FDAAs 24
4. Applications of FDAAs and Their Derivatives..........................34
   4.1. Tools for Studies of PG Synthesis Pattern and
        Dynamics of PG 34
   4.2. PG Recycling/Remodelling in *Streptococcus
        pneumoniae* 36
   4.3. FDAAs Reveal PG in Strains not Previously Known
        to Possess It 38
   Conclusions 41
   References..42

CHAPTER 2 Time-Lapse Microscopy and Image Analysis of *Escherichia coli* Cells in Mother Machines 49
Y. Yang, X. Song, A.B. Lindner

1. Introduction 49
2. Experimental Designs 50
   2.1. General Principles of the Mother Machine 50
   2.2. Device Design Considerations 51
   2.3. Time-Lapse Considerations 53
3. Experimental Procedures 54
   3.1. Making Polydimethylsiloxane Mother Machine Devices 55
   3.2. Setting up the Fluidic System 55
   3.3. Cell Culture and Loading 56
   3.4. Microfluidic and Time-Lapse Setup 57
4. Image Analysis for Lineage Construction and Single-cell Traits 57
   4.1. Preprocessing 57
   4.2. Segmentation Approaches 58
   4.3. Lineage Approaches 61
   4.4. Image Analysis Performances and Results 64
   4.5. Image Analysis Summary 66
   Acknowledgment 66
   References 66

CHAPTER 3 Microfluidics for Bacterial Imaging 69
L.E. Eland, A. Wipat, S. Lee, S. Park, L.J. Wu

1. Introduction 69
2. Fabrication of Microfluidic Devices 71
   2.1. Design 72
   2.2. Production of Silicon Wafers 74
   2.3. Production of the Microfluidics Chip 77
3. Fluid Flow 77
4. Additional Considerations When Designing and Setting Up a Microfluidics System 79
5. Computational Analysis and Control of Bacterial Microfluidic Systems 80
6. Microfluidics for the Engineering of Bacterial Systems 83
7. Biofilm, Microbial Ecology and species–species Interactions 86
   7.1. Multispecies Biofilm Microfluidics 87
   7.2. Spatial Arrangements and Interactions 90

8. Cell Cycle Analysis and Size Homeostasis Studies 90
9. Cell Shape and Geometry Study 94
10. Future Perspectives 96
11. Methods 97
    Acknowledgements 105
    References 105

SECTION 2   ELECTRON MICROSCOPY

CHAPTER 4   Electron Cryotomography 115
C.M. Oikonomou, M.T. Swulius, A. Briegel, M. Beeby,
Q. Yao, G.J. Jensen

1. Introduction to Electron Cryotomography Technology 115
2. Applications of ECT to Microbiology 116
3. Comparison to Other Techniques 120
   3.1. Limitations of ECT 120
   3.2. Advantages of ECT 129
   Conclusions and Future Directions 133
   Acknowledgements 133
   References 133

Index 141

Contributors

M. Beeby
Imperial College of London, London, United Kingdom

A. Briegel
Institute of Biology, Leiden University, Leiden, The Netherlands

L.E. Eland
Interdisciplinary Computing and Complex BioSystems Research Group, School of Computing Sciences; Centre for Bacterial Cell Biology, Institute for Cell and Molecular Biosciences, Newcastle University, Newcastle upon Tyne, Tyne and Wear, United Kingdom

Y.-P. Hsu
Indiana University, Bloomington, IN, United States

G.J. Jensen
California Institute of Technology; Howard Hughes Medical Institute, Pasadena, CA, United States

S. Lee
Centre for Bacterial Cell Biology, Institute for Cell and Molecular Biosciences, Newcastle University, Newcastle upon Tyne, Tyne and Wear, United Kingdom

A.B. Lindner
INSERM, U1001; Faculté de Médecine, Université Paris Descartes; Center for Research and Interdisciplinarity, Paris, France

X. Meng
Indiana University, Bloomington, IN, United States

C.M. Oikonomou
California Institute of Technology, Pasadena, CA, United States

S. Park
Interdisciplinary Computing and Complex BioSystems Research Group, School of Computing Sciences; Centre for Bacterial Cell Biology, Institute for Cell and Molecular Biosciences, Newcastle University, Newcastle upon Tyne, Tyne and Wear, United Kingdom

X. Song
INSERM, U1001; Faculté de Médecine, Université Paris Descartes, Paris, France

M.T. Swulius
California Institute of Technology, Pasadena, CA, United States

M.S. VanNieuwenhze
Indiana University, Bloomington, IN, United States

A. Wipat
Interdisciplinary Computing and Complex BioSystems Research Group, School of Computing Sciences; Centre for Bacterial Cell Biology, Institute for Cell and

Molecular Biosciences, Newcastle University, Newcastle upon Tyne, Tyne and Wear, United Kingdom

L.J. Wu
Centre for Bacterial Cell Biology, Institute for Cell and Molecular Biosciences, Newcastle University, Newcastle upon Tyne, Tyne and Wear, United Kingdom

Y. Yang
INSERM, U1001; Faculté de Médecine, Université Paris Descartes; Center for Research and Interdisciplinarity, Paris, France

Q. Yao
California Institute of Technology, Pasadena, CA, United States

Preface

Two of our favorite quotes are "It is very easy to answer many of these fundamental biological questions; you just look at the thing!" (Richard Feynmann)[a] and "Sometimes you can see a lot just by looking" (often attributed to Yogi Berra). Both are wonderfully true—in fact the history of cell biology has been punctuated by advances in microscopy and imaging that have allowed scientists to see ever more about the inner workings of the cell.

This volume of *Methods in Microbiology* is therefore dedicated to current and emerging technologies in the field of microbial imaging. The intended audience includes research scientists as well as medical and industrial professionals interested in microbes and the technologies used to understand them. We thank the authors of each chapter and Alex White, Hannah Colford, Sarah Lay, and Vignesh Tamilselvvan at Elsevier for overseeing the project.

Colin Harwood
Grant J. Jensen

---

[a]There's plenty of room at the bottom: An invitation to enter a new field of physics, a talk to the American Physical Society Dec. 1959 published in Caltech Engineering and Science 23:22–36.

SECTION 1

Fluorescent Microscopy

CHAPTER 1

Methods for visualization of peptidoglycan biosynthesis

Y.-P. Hsu[1], X. Meng[1], M.S. VanNieuwenhze[2]

Indiana University, Bloomington, IN, United States
[2]*Corresponding author: e-mail address: mvannieu@indiana.edu*

Peptidoglycan (PG) is a rigid macropolymer surrounding the cytoplasmic membrane of bacteria. It is an essential cell wall component that provides bacterial cells with the mechanical strength to withstand turgor pressure and environmental stress. Bacteria have evolved a sophisticated machinery to synthesize, and to maintain the integrity of, peptidoglycan. This machinery includes proteins responsible for PG synthesis, modification, and degradation, which precisely regulate PG growth, structure, and cell shape during the cell cycle. PG synthesis is also coordinated with the bacterium's cytoskeletal system in order to coordinate cell elongation, division, and, where applicable, sporulation. Disruption of this precisely regulated machinery results in defective cell growth and can easily lead to cell lysis. Due to its essential nature, scientists from various fields have expended great efforts towards the study of the PG biosynthetic pathway and its structure.

Since PG is essential for most bacterial species, it has been regarded as a key target for antimicrobial chemotherapy. To illustrate this point, the β-lactam antibiotics, perhaps the most common class of antibiotics used to treat bacterial infections, manifest their antibiotic activity through the inhibition of the PG biosynthetic pathway (specifically, the transpeptidation step). Antibiotics targeting this pathway also have the benefit of limited toxicity since there is no eukaryotic equivalent to the PG biosynthesis pathway. However, the increasing use of antibiotics has also led to the increasing prevalence of drug resistance, which has begun to compromise the efficacy of the current antibiotic arsenal. In 2013, the Centers for Disease Control and Prevention of the United States highlighted antibiotic resistance as an urgent threat and emphasized the critical need for development of new antibiotics and identification of new antibiotic targets. As this mandate pertains to agents that target the PG biosynthetic pathway, a more thorough understanding of the PG biosynthesis and growth dynamics is required.

Historically, scientists have determined the chemical composition of PG via enzymatic/chemical degradation followed by purification of the degradation products by high-performance liquid chromatography (HPLC) and product characterization by

---

[1]These authors contributed equally.

mass spectrometry (MS). However, our knowledge of the details of PG architecture in bacterial cells, and the coordination between PG synthesis and the relevant protein machineries that coordinate this process, remains limited. While advances in microscopy have significantly advanced the ability to study peptidoglycan synthesis and architecture, the full impact of these advances have not been realized due to the lack of probes to enhance our ability to visualize peptidoglycan synthesis and dynamics in real time and/or in live bacterial cells.

The goal of this chapter is to review the various imaging strategies that have been developed for studying PG biosynthesis and dynamics. It begins with a brief introduction to PG structure, and the enzymes involved in the PG biosynthetic pathway, which provides readers with the background information required for a clearer understanding of the various imaging strategies employed. We then discuss imaging techniques based on electron microscopy (EM), followed by imaging techniques that utilize fluorescence microscopy (FM). We survey chemical probes that are used to label PG and discuss their applications. Finally, we present the characterization and applications of newly developed metabolic probes that have significantly advanced the toolset for PG research.

1 PEPTIDOGLYCAN STRUCTURE AND BIOSYNTHESIS
1.1 PEPTIDOGLYCAN STRUCTURE AND CONFIGURATION

Peptidoglycan is composed of glycan strands cross-linked by peptide chains (Fig. 1) (Typas, Banzhaf, Gross, & Vollmer, 2012; Vollmer, Blanot, & de Pedro, 2008). The glycan strands are made of alternating N-acetylmuramic acid (MurNAc) and N-acetylglucosamine (GlcNAc) residues that are connected through β-1,4-glycosidic linkages. Attached to each MurNAc residue is a pentapeptide chain (also known as the stem peptide), appended to a C(3)-lactate anchor, that contains both D- and L-amino acids. In *Escherichia coli* and most Gram-negative species, the peptide chain consists of (in order) L-Ala, γ-D-Glu, *meso*-Dap, and D-Ala-D-Ala. However, in most Gram-positive species, *meso*-Dap is replaced with an L-Lys. The stem peptide chains are subsequently cross-linked, either directly or through a short peptide bridge that results in the typical "mesh-like" structure. The cross-linking reaction significantly enhances the mechanical strength of PG to protect cells from environmental stress and to resist lysis. Even after cross-linking, PG remains a highly elastic material since the bonds between glycan strands, peptide chains, and cross-links are flexible. This flexibility allows bacteria to change morphology in response to the environmental factors.

Gram-negative cells have a single layer of PG with a thickness of about 5 nm (Yao, Jericho, Pink, & Beveridge, 1999). This PG layer is sandwiched by inner and outer membranes. In Gram-positive bacteria, where there is no outer membrane, PG comprises the outermost shell of the cell. This PG layer is much thicker (20–50 nm) than the PG layer in Gram-negative organisms (Vollmer et al., 2008). Because they lack an outer membrane, the PG in Gram-positive organisms is exposed to the surroundings, generally making Gram-positive organisms more sensitive to antibiotics targeting PG biosynthesis.

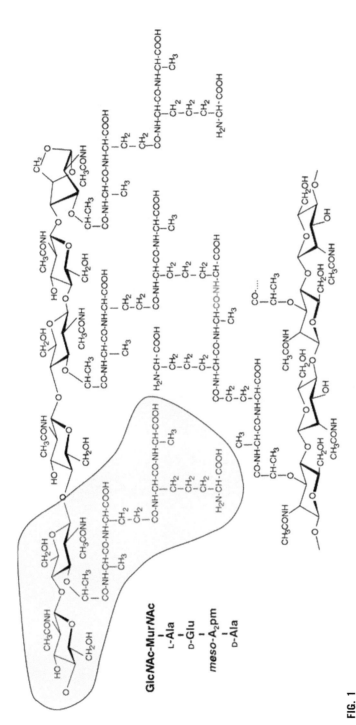

FIG. 1

Peptidoglycan structure in *E. coli*. The area outlined in *yellow* illustrates the monomeric PG subunit. The terminal D-Ala residue at the fifth position of the stem peptide is usually removed in mature *E. coli* PG (Vollmer et al., 2008).

1.2 ENZYMATIC PATHWAY OF PEPTIDOGLYCAN BIOSYNTHESIS

PG synthesis occurs in three overall stages at three different locations:

(1) Synthesis of the Park's nucleotide in the cytoplasm
(2) Construction of lipid-anchored precursors (lipid I and lipid II) at the cytoplasmic membrane
(3) Insertion of lipid II into existing PG sacculi in periplasm.

1.2.1 Formation of the Park's nucleotide

The Park's nucleotide is a PG precursor composed of a UDP-MurNAc residue and a pentapeptide chain (van Heijenoort, 2007). Its synthesis is initiated by an enolpyruvyl transferase, MurA, which couples a phosphoenolpyruvate to UDP-N-acetylglucosamine (GlcNAc) to give an UDP-GlcNAc-EP (Kock, Gerth, & Hecker, 2004). This intermediate is then reduced by MurB, a NADPH-dependent reductase, to introduce the C(3) lactic acid anchor for the stem peptide. This reaction is then followed by sequential coupling of amino acids through a series of ATP-dependent ligases (MurC-F) that generate the pentapeptide chain. For example, in *E. coli*, this process includes L-Ala incorporation into UDP-MurNAc (by MurC), the coupling of D-Glu (by MurD), *meso*-Dap (by MurE), and finally D-Ala-D-Ala dipeptide (by MurF) (Baum et al., 2009; Bratkovic, Lunder, Urleb, & Strukelj, 2008; Mol et al., 2003; Real & Henriques, 2006; Sink et al., 2013).

1.2.2 Synthesis of lipid II

Lipid II is derived from the Park's nucleotide and is the final monomeric precursor utilized for incorporation into existing PG (de Kruijff, van Dam, & Breukink, 2008). It comprises a disaccharide-pentapeptide linked to a pyrophosphate-tethered lipid anchor. The synthesis of lipid II starts by attaching Park's nucleotide to a lipid carrier, undecaprenyl phosphate (C_{55}-P), in a phosphoryl transfer reaction mediated by MraY, to give the first lipid-linked intermediate, lipid I (Bouhss, Crouvoisier, Blanot, & Mengin-Lecreulx, 2004). Lipid I is then coupled with a UDP-GlcNAc residue, through a reaction mediated by MurG, to provide lipid II (Mengin-Lecreulx, Texier, Rousseau, & van Heijenoort, 1991). Because lipid II is embedded in the inner leaflet of the cytoplasmic membrane, flippases are required to transfer lipid II across the membrane into the periplasm where PG synthesis takes place.

1.2.3 Nascent PG insertion and PG remodelling

Insertion of lipid II into existing PG is mainly conducted by PG synthases, the so-called penicillin-binding proteins (PBPs) (Sauvage, Kerff, Terrak, Ayala, & Charlier, 2008). Once lipid II is flipped to the periplasmic side of the cell membrane, PBPs perform the three functions required for PG synthesis: (1) glycosyltransferases (GTases) lengthen the glycan strands through the addition of the lipid II disaccharide to the reducing end of the strand (Lairson, Henrissat, Davies, & Withers, 2008); (2) transpeptidases (TPases) construct 4–3 cross-links (e.g. D-Ala to *meso*-Dap) between stem peptide chains to provide the "mesh-like" structure typical of PG (Lebar et al., 2013);

and (3) carboxypeptidases (CPases) cleave the D-Ala at the fifth position of pentapeptide chains for PG modification (Ghosh, Chowdhury, & Nelson, 2008).

In addition to PBPs, other enzymes are also involved in the direct modification of PG for bacterial growth. L,D-Transpeptidases (Ldts) also catalyse the formation of cross-links between glycan strands (Mainardi et al., 2005). Ldts construct 3–3 cross-links between the *meso*-Dap residues of peptide stems. It is known that Ldts are not inhibited by most penicillin derivatives (Lecoq et al., 2012). The frequency of 3–3 cross-links in Gram-negative bacteria is higher than that in Gram-positive bacteria (Pisabarro, de Pedro, & Vazquez, 1985).

PG hydrolases, known as autolysins, are also required for cell growth. They are responsible for modifying existing PG to facilitate the insertion of new PG and cell wall constriction during cytokinesis (Blackman, Smith, & Foster, 1998). Examples include amidases, which cleave peptide chains from glycan strands, and endopeptidases, which cleave peptide bonds.

1.3 STRUCTURAL VARIATIONS IN PEPTIDOGLYCAN STRUCTURE

PG is essential in most bacteria, and its features described in the preceding paragraphs are highly conserved across bacterial species. However, structural variations have been found in the glycan strands, the peptide chains, and the nature or type of the interpeptide cross-link (Vollmer et al., 2008).

One example of a structural variation within a glycan strand can be found at its reducing end after cleavage by lytic transglycosylases. In *Staphylococcus aureus*, the reducing end could be either a MurNAc or a GlcNAc residue, depending on the enzyme involved in the remodelling reaction (Boneca, Huang, Gage, & Tomasz, 2000). In some Gram-negative species, however, neither MurNAc nor GlcNAc is found. Instead, it is a cyclized MurNAc derivative, called 1,6-anhydroMurNAc, which terminates the glycan strands. The existence of 1,6-anhydroMurNAc is a hallmark of glycan strand ends in these species and has been used to determine glycan strand length using HPLC in combination with MS (König, Claus, & Varma, 2010).

The amino acid composition in PG peptide chains may vary between bacterial species. In most species, the first amino acid in peptide chains is L-Ala. However, Gly and L-Ser are encountered in some species, such as *Mycobacterium leprae* and *Butyribacterium rettgeri* (Table 1) (Mahapatra, Crick, & Brennan, 2000). This probably results from the low specificity of MurC towards L-Ala, Gly, and L-Ser in these species (see Fig. 2) and the concentration of each in the respective growth environments.

The second amino acid in PG peptide chains, D-γ-Glu, is relatively conserved across all bacterial species. Modifications of the D-γ-Glu residue are known in several bacterial species, but these modifications arise from reactions subsequent to its incorporation into the stem peptide. The greatest variation, on the other hand, is found at the third position. MurE catalyses the ligation of the third amino acid into the growing stem peptide. In most Gram-negative species, and some Gram-positives, this amino acid is *meso*-Dap. In most Gram-positive organisms, this amino acid is

Table 1 Amino Acid Variations in PG Peptide Chain (Vollmer et al., 2008)

Position	Residue Encountered	Examples
1	L-Ala	Most species
	Gly	*Mycobacterium leprae, Brevibacterium imperiale*
	L-Ser	*Butyribacterium rettgeri*
2	D-Isoglutamate	Most Gram-negative species
	D-Isoglutamine[a]	Most Gram-positive species, Mycobacteria
	threo-3-Hydroxyglutamate[a]	*Microbacterium lacticum*
3	*meso*-A_2pm	Most Gram-negative species, Bacilli, Mycobacteria
	L-Lys	Most Gram-positive species
	L-Orn	Spirochetes, *Thermus thermophilus*
	L-Lys/L-Orn	*Bifidobacterium globosum*
	L-Lys/D-Lys	*Thermotoga maritima*
	LL-A_2pm	*Streptomyces albus, Propionibacterium petersonii*
	meso-Lanthionine	*Fusobacterium nucleatum*
	L-2,4-Diaminobutyrate	*Corynebacterium aquaticum*
	L-Homoserine	*Corynebacterium poinsettiae*
	L-Ala	*Erysipelothrix rhusiopathiae*
	L-Glu	*Arthrobacter* J. 39
	Amidated *meso*-A_2pm[a]	*Bacillus subtilis*
	2,6-Diamino-3-hydroxypimelate[b]	*Ampuraliella regularis*
	L-5-Hydroxylysine[b]	*Streptococcus pyogenes*[c]
	N^γ-Acetyl-L-diaminobutyrate[a]	*Corynebacterium insidiosum*
4	D-Ala	All bacteria
5	D-Ala	Most bacteria
	D-Ser	*Enterococcus gallinarum*
	D-Lac	*Lactobacillus casei*, Enterococci with acquired resistance to vancomycin

[a]These residues result from the modification posterior to the Mur ligase reactions.
[b]The formation mechanism of these residues is unclear.
[c]A 10:1 ratio of lysine to hydroxylysine was found in *Streptococcus pyogenes*.

L-Lys. Other amino acid variations at this site have been encountered, including *meso*-lanthionine, L-Orn, and L-Glu (Schleifer & Kandler, 1972).

The amino acids at the fourth and fifth positions are incorporated into PG as a dipeptide, specifically, D-Ala-D-Ala (Duncan, van Heijenoort, & Walsh, 1990). Synthesis of the D-Ala-D-Ala dipeptide is facilitated by D-Ala-D-Ala ligase (Ddl). The dipeptide is subsequently incorporated into the stem peptide by MurF, as described earlier (Fig. 2). The structure of this dipeptide is highly conserved. D-Ala is predominantly found at the fourth position in all bacteria known to date, while D-Lactate and

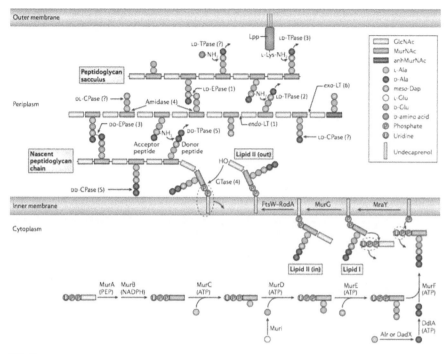

FIG. 2

Enzymatic pathway for PG biosynthesis and the proteins involved in PG remodelling in *E. coli* (Typas et al., 2012).

D-Ser may be found, albeit rarely, in the fifth position in some species (Cetinkaya, Falk, & Mayhall, 2000).

In addition to the stem peptide, the structure of cross-linkages may also vary. In *E. coli* and *Bacillus subtilis*, cross-linkages are installed directly without any additional bridging amino acids. In contrast, *S. aureus* utilizes a bridge link composed of five glycine residues attached to the terminal amino group of the lysine residue. Other organisms may also employ various bridge links (Vollmer et al., 2008).

Scientists have been attempting to develop methods for PG-specific analysis based on the unique structure of PG. In the following sections, we will discuss the labelling strategies that have been commonly employed for PG imaging.

2 STRATEGIES FOR PEPTIDOGLYCAN IMAGING: ELECTRON MICROSCOPY

2.1 REVEALING BACTERIAL MORPHOLOGY

The unique chemical composition of PG, as well as its prominent role in various aspects of the bacterial cell cycle, has prompted scientists to study its synthesis and dynamics. For example, efforts have been directed at learning how bacteria coordinate PG synthesis with cellular growth and how PG synthesis may be modified as bacteria respond to their growth environment. In order to execute such studies, it is crucial to be able to visualize the fine structure of PG.

In 1675, Antony van Leeuwenhoek reported the first descriptions of microbial morphology (e.g. spirals, rods, etc.) using microscopy. However, it was not until 300 years later that high-resolution observations of cell morphology were reported by Mudd et al. using electron EM (Mudd, Polevitzky, Anderson, & Chambers, 1941; Mudd, Polevitzky, Anderson, & Kast, 1942; Stuart Mudd, 1941) (Fig. 3).

EM is a powerful and commonly used technology for microstructural studies. It constructs high-resolution images by casting accelerated electron beams onto the specimens and then collecting the signals of transmitted electrons (transmission electron microscopy, TEM) (Williams & Carter, 1996), reflected electrons (reflection electron microscopy) (Yagi, 1987), or secondary electrons (scanning electron microscopy, SEM) (Vernon-Parry, 2000). The resolution of EM is determined by the wavelength of the incident electron beam. Current high-voltage EM can achieve subnanoscale illumination where resolutions on the order of about 2 Å can be achieved (Bartesaghi et al., 2015). The signal intensity of TEM is mainly dependent on the electron density of the specimens. Namely, objectives having high electron density give a strong signal (usually seen in dark colour), while those having low electron density have less signal and might be invisible in EM images. Therefore, appropriate staining, using heavy atoms, is required for the specimens containing only low electron density elements, such as peptidoglycan.

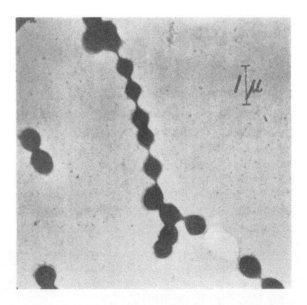

FIG. 3

EM image of streptococcal cells reported by Mudd et al. Magnification: ×12,000 (Mudd et al., 1941).

2.2 ELUCIDATING PG STRUCTURE VIA EM

In early EM studies, scientists developed peptidoglycan-specific staining methods in order to investigate PG structure and architecture. These staining methods relied upon high binding affinity between PG structure and the probe molecules. Formanek et al., in 1970, reported the use of mercury-containing compounds, (3-carboxy-2-hydroxy-5-sulfophenyl)(hydroxy)mercury and anhydro-3-hydroxymercuri-4-toluenesulfonic acid, to stain isolated PG from *Spirillum serpens* (Formanek & Formanek, 1970). Each of these compounds forms covalent bonds with free amine groups displayed on PG, allowing excess stain to be removed through washing steps. The high electron density of mercury allows PG to be observed via EM (Fig. 4). PG structure stained by mercury-containing compounds appeared darker in colour in the image, while the background was lighter. This image revealed that the isolated PG from *S. serpens* maintained the morphology of the cell.

EM has also been employed to study PG synthesis. Cell elongation and division are essential events of bacterial life cycle. PG formation is required for the construction of a new lateral cell wall for cell elongation as well as a transverse cell wall, the septum, for cell division. The enzymatic events associated with new PG synthesis/insertion and old PG remodelling are highly sophisticated and are precisely controlled by the bacterial protein machinery. Visualizing where and when new PG is inserted into existing PG is important for a clearer understanding of bacterial

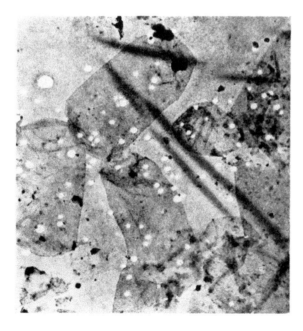

FIG. 4

EM images of isolated *Spirillum serpens* PG stained by mercury-containing PG-specific compounds. Magnification ×32,000 (Formanek & Formanek, 1970).

physiology. To achieve this goal, a significant effort was directed towards the development of methods to differentiate new PG from old PG using microscopy. In 1997, de Pedro et al., used an immunostaining approach to investigate the insertion of new PG in *E. coli* during its life cycle (de Pedro, Quintela, Holtje, & Schwarz, 1997). This work took advantage of the ability of bacterial cells to incorporate D-cysteine into peptidoglycan in place of D-alanine. After isolation purified sacculi, the cysteine-decorated PG was treated with a thiol-reactive biotinylating reagent (biotin-HPDH) that introduced a tag for capture with an antibiotin antibody (rabbit antibiotin antiserum), followed by protein A-10 nm gold immunolabelling.

Fig. 5 shows the immunoelectron microscopy images of isolated *E. coli* PG from de Pedro et al. Cells incubated in the presence of D-cysteine revealed a homogeneous distribution of gold labelling throughout the cells (Fig. 5A). Because the incorporation of D-cysteine is dependent upon the PG synthesis activity (through PBPs or Ldts), the EM signal was PG specific and growth dependent. To further reveal the insertion sites of new PG, they did pulse-and-chase experiments where D-cysteine-decorated cells were transferred to D-cysteine-free medium for additional incubation so that old PG was gold decorated but new PG was not. Their results showed that chased cells had a homogeneous yet diluted distribution of old PG lateral wall cell surface, which confirmed the model of diffuse insertion of PG in *E. coli* (Fig. 5B). Also, they observed a high density of EM signal at the cell poles, suggesting that the PG synthesis activity remained relatively inert at poles during cell growth (Fig. 5C). On the other hand, the

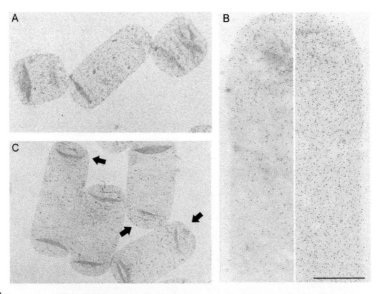

FIG. 5

Immunoelectron microscopy images of *E. coli*. (A) D-Cysteine/gold decorated PG, (B) comparison between PG with and without D-cysteine-free chasing, and (C) D-cysteine/gold chased PG with D-cysteine-free chasing. Scale bar: 1 μm (de Pedro et al., 1997).

cells had the lowest EM signal at the midcell, the perspective septum synthesis site of *E. coli*, indicating the elevated PG synthesis activity in preparation for cell division.

Immunostaining provides a powerful approach for studying PG synthesis and dynamics because it enables PG-specific labelling. This approach can also be adapted for FM by decorating the staining reagent with fluorophores instead of heavy elements (de Pedro et al., 1997; Tiyanont et al., 2006). For example, the EM results reported by de Pedro et al. have been further confirmed using FM. This was achieved by replacing gold-modified protein A with fluorescent goat antirabbit antiserum to fluorescently label the D-cysteine decorated PG.

2.3 CRYO-TEM: STUDYING CELL WALL ORGANIZATION IN NATURAL STATE

Recent advances in EM have made it a powerful tool to illuminate PG morphology and structure with enhanced resolution. However, EM images should be interpreted with caution because the preparation of the sample (e.g. cell fixation and staining) might disrupt the PG structure (Chao & Zhang, 2011; Vollmer et al., 2008). It has been known that chemical fixation may cause cells to shrink, which can lead to incorrect measurement of cell size using EM images. While efforts have been directed at optimizing fixation conditions and to study their impact on cell morphology (Chao & Zhang, 2011), the introduction of cryo-TEM has provided a method to circumvent potential issues associated with fixation. Cryo-TEM illuminates specimens at cryogenic temperatures (typically in a liquid nitrogen environment) (Chao & Zhang, 2011; Kühlbrandt, 2014). It allows the observation of specimens without fixation and staining. Cryo-EM specimens are preserved in the frozen-hydrated state by rapid freezing using liquid nitrogen. It preserves the components and morphology of the specimens while retaining their structures in a native environment.

Matias reported *E. coli* section images captured by cryo-TEM (Matias, Al-Amoudi, Dubochet, & Beveridge, 2003; Matias & Beveridge, 2005). The cells were rapidly frozen using high-pressure freezing in the presence of cryoprotectant to ensure consistent vitrification of the specimen. The specimen was then sectioned by ultramicrotome to give a 50-nm-thick specimen slice and subsequently imaged. As shown in Fig. 6, the cryo-TEM images of *E. coli* section revealed the organization of bacterial cell wall without fixation and staining. The relative position of outer membrane, PG, and inner membrane (or plasma membrane, PM) could be observed clearly. The image nicely captured the natural state of cell wall organization since, if the vitrified specimens were thawed, more than 70% cells regained viability. The outer and inner membranes, which are composed of densely packed lipid phosphates, were observed remarkably as dark-colour envelopes surrounding the cell. In contrast, the structure of PG is less noticeable since it is mainly composed of low electron density elements (e.g. H, C, N, O), which do not contribute greatly to the EM signal. Nevertheless, the compact PG structure still had good EM signal and could be clearly distinguished from the background. PG was observed as a thin band located between outer and inner membranes.

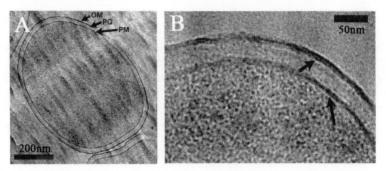

FIG. 6

(A) Cryo-TEM image of E. coli sectioning. (B) Zoom-in image of E. coli cell wall organization (Matias et al., 2003).

The E. coli images displayed in Fig. 6 suffered from sectioning compression; the tensional stress resulting from sectioning remained in the specimen and led to the oblong shape of the cells. However, this artefact could be eliminated because it was a systematic error that could be measured and predicted. Analysis of the cryo-TEM images revealed a thickness of 6.35 nm for E. coli PG (Matias et al., 2003). This value is highly consistent with the results measured by atomic force microscopy (AFM) (Yao et al., 1999).

2.4 THREE-DIMENSIONAL EM TECHNIQUE: CRYOELECTRON TOMOGRAPHY

As introduced earlier, TEM constructs images by casting an accelerated electron beam through a specimen and constructing two-dimensional images from the interaction of the electrons with the specimen. However, biological macromolecules and cells are highly sophisticated three-dimensional architectures. Extracting the information from the third dimension of the structure is crucial for understanding the spatiotemporal relationship between macromolecules. A recently developed cryo-TEM application, called cryoelectron tomography (CET), has effectively addressed this limitation and significantly impacted the field of microstructure imaging. CET is a specialized application of TEM where the specimens are imaged (Fig. 7A) as they are tilted relative to the incident electron beam (Lucic, Rigort, & Baumeister, 2013; McIntosh, Nicastro, & Mastronarde, 2005; Murphy, Leadbetter, & Jensen, 2006). Sequential tilting of the specimens generates a series of two-dimensional images that can be further reconstructed into three-dimensional structures. CET can thus visualize microstructures, such as PG, in three dimensions with high resolution (Fig. 7B).

The relative architecture of PG in bacterial cells has remained a long-standing question in microbiology. Two models proposed have been proposed based on experimental observations. The layered model proposes that glycan strands lie on the surface of the cell wall, enveloping the amorphous cell body. This model was based on the observation of cross-section images where gaps between glycan strands

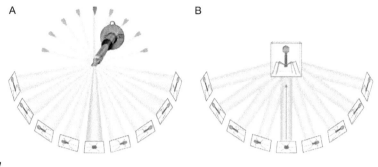

FIG. 7

Scheme illustrating the concept of cryoelectron tomography. A specimen in the TEM chamber is imaged as it is tilted at different angles (McIntosh et al., 2005).

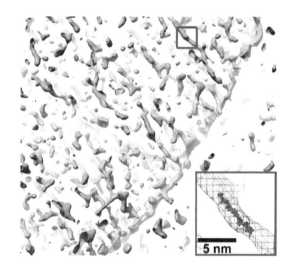

FIG. 8

C. crescentus PG structure constructed by cryo-electron tomography. The *green-shaded region* demarcates peptidoglycan. The putative glycan strand is boxed in *blue*. Noise densities outside of the PG are coloured *grey* for clarity (Gan et al., 2008).

appeared to be perpendicular to the surface of the cell (De Petris, 1967; Verwer, Nanninga, Keck, & Schwarz, 1978). Alternatively, the scaffold model proposed that PG glycan strands are oriented perpendicular to the surface of the cell (Dmitriev et al., 2003). To gain insight into this question, Gan investigated the three-dimensional configuration of *Caulobacter crescentus* PG using CET (Gan, Chen, & Jensen, 2008). As shown in Fig. 8, their results revealed that thin tubes were oriented parallel to the surface of the isolated PG and roughly perpendicular to the long axis of the cell. These thin tubes were inferred as PG glycan strands because (1) glycan strands have higher electron density due to its carbohydrate backbone and (2) many tubes were longer than a peptide cross-link. Their remarkable uniformity in thickness also suggested that PG glycan strands exist as single strands rather than bundles. Based on these results, Gan et al. proposed that PG has a poorly ordered configuration where the majority of glycan strands formed a single layer that lies along the cell surface.

EM has greatly advanced our knowledge of PG structure and architecture. EM has been employed to illuminate isolated PG structures using immunostaining and, later, the organization of entire cell wall in its natural state at cryogenic temperatures. The development of high-voltage EM significantly improved EM resolution to subnanoscale, which enabled the molecular level construction of biological architectures. Finally, the use of CET provided three-dimensional information for the analysis of PG architecture.

2.5 MORE TOOLS: AFM IN BACTERIAL CELL WALL STUDIES

In parallel to EM, the development of AFM has also significantly impacted the field of microstructure imaging. AFM was invented in the late 1900s by Binnig, Rohrer, Gerber, and Weibel (1982). This technique utilizes a cantilever with an ultra-thin tip to scan the specimen surface. When the tip is brought to the surface of the specimen, force generated from the tip–specimen interaction is reflected to the cantilever, which is transformed into optical signal and analysed by the computer (Fig. 9) (Allison et al., 2010; Ikai, 2010). AFM can be applied to detect mechanical forces (direct contact with the surface), van der Waals forces, chemical bondings, electrostatic forces, magnetic forces, etc. Based on the experimental design, AFM could have several advantages over TEM for studying biological specimens (Yao et al., 1999). First of all, AFM provides exact Z-axis measurement because it generates specimen images based on its thickness (or height). Second, AFM specimens are

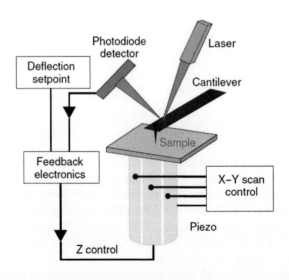

FIG. 9

A schematic depiction of atomic force microscopy (Allison, Mortensen, Sullivan, & Doktycz, 2010).

illuminated under mild conditions compared to TEM where specimens are lyophilized in a vacuum chamber. AFM can even analyse specimens in water and buffer, which allows biological specimens to remain in a hydrated state. Third, because the contact between cantilever tip and specimens can be adjusted, AFM can also be used to measure the rigidity and elasticity of specimens, enabling analysis of the surface properties of bacterial cell wall and PG (Yao et al., 1999).

Hayhurst reported a remarkable example of the application of AFM. This study investigated the shape and structure of purified PG glycan strands (Hayhurst, Kailas, Hobbs, & Foster, 2008). The purified glycan strands were prepared by treating isolated PG with an amidase (*S. aureus* Atl.) to remove cross-links, followed by purification using ion exchange chromatography. AFM images showed that the length of single PG glycan strand in *B. subtilis* ranged from 0.2 to 5 μm with an average of 1.3 μm. This result was surprising because it suggested that a glycan strand could be longer than a *B. subtilis* cell. This finding also suggested that glycan strands must be "wrapped" in some fashion in order to fit onto the cell surface.

In order to understand how the glycan strands fit into the overall PG structure, Hayhurst also investigated the inner surface of isolated PG by AFM (Hayhurst et al., 2008). The inner surface of PG provides a better platform for studying PG architecture than the outer surface because environmental stress may result in the structural damage on the cell surface. They found that the inner PG surface was packed with cable-like materials with an average width about 50 nm. These "cables" were parallel to the short axis of the cell and composed of helically arranged fibers that could be observed under high-resolution AFM. Based on this observation, Hayhurst proposed that, during PG synthesis, multiple glycan strands are polymerized into PG fibers. These fibers are coiled in a helical manner to give a "cable" with a width on the order of 50 nm. The cable is then inserted into existing PG by cross-linking between cables. Their model provides an experimental explanation of the glycan strand length. It also provided an insight into a potential mechanism for resisting cell lysis since wrapping into "cables" may be expected to significantly increase the overall strength of the macroscopic PG structure (Figs. 10 and 11).

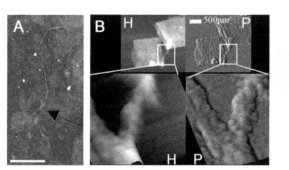

FIG. 10

AFM images of (A) purified PG glycan strands (noted by *black arrow*, scale bar: 1 μm); (B) isolated PG from gently broken cells. *H*, height imaging; *P*, phase imaging. Insertion shows an interpretative diagram of PG "cables" observed in (B) (Hayhurst et al., 2008).

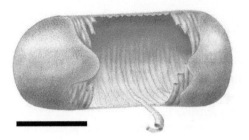

FIG. 11

Model of *B. subtilis* PG architecture proposed by Hayhurst et al. (2008).

3 STRATEGIES OF PEPTIDOGLYCAN IMAGING: OPTICAL MICROSCOPY

3.1 OPTICAL MICROSCOPY OPERATION AND APPLICATION

Optical microscopy, or light microscopy, is an important technique for studying microscale structures in various disciplines. Optical microscopy has been heavily employed in biological studies because it enables illumination of specimens under relatively mild conditions. Instead of using an accelerated electron beam as the signal source, optical microscopy utilizes photons as signal source. Furthermore, it can generate images at normal temperature and pressure. The mild conditions for image collection enable time-lapse studies that, in principle, allow monitoring of a specimen over time without having to resort to fixing.

An optical microscope is composed of a light source, a condenser lens, an objective lens, and an eyepiece (ocular lens) (Davidson & Abramowitz, 2002; Mertz, 2010). Photons generated from a light source are aligned and focused onto a specimen by the condenser lens. The deviated light passing through the specimen is then collected by objective lenses, redirected into eyepiece lens, and eventually imaged on the retina of user's eyes.

Optical microscopy generates images based on the absorbance of the specimens. Namely, specimens having thick and compact structure result in better illumination because less incident light can pass through them and be detected by the observer. On the other hand, specimens such as prokaryotic and eukaryotic cells have an attenuated ability to absorb light and specific features may not be observed in optical microscopy. As a result, staining is frequently used to enhance visualization of such specimens.

Microscopy techniques have also been developed to illuminate transparent samples without staining. Phase contrast microscopy (PCM) is one of the most popular techniques used to illuminate unstained specimens (Davidson & Abramowitz, 2002; Zernike, 1942). In the early 1930s, Zernike et al., discovered that a phase shift of approximately 1/4 wavelength occurs when the incident light passes through a specimen. This phase difference between the deviated light and direct light (the incident light that does not interact with the specimen) cannot be directly observed. However, Zernike successfully devised a method to amplify the phase difference to 1/2 wavelength by inserting a phase ring and phase plate into the light path of the microscope. As a result, the deviated light and direct light arriving at the same focusing plane of the eyepiece generate destructive interference, giving a darker image of the specimen against a lighter background (Fig. 12A). The development of PCM enabled visualization of specimens without the need for staining. It has been applied to study the biological specimens, including bacterial cells and isolated bacterial peptidoglycan.

Another strategy commonly used to enable visualization of unstained specimens is called differential interference contrast (DIC) microscopy (Davidson & Abramowitz, 2002; Murphy & Davidson, 2012). DIC illuminates a specimen based on the principle of interferometry. Incident light is first polarized using a polarizer and then separated into two beams by a prism. These two beams are spatially displaced (sheared) and vibrating in perpendicular directions. When they enter the condenser, they are focused on the specimen plane and then pass through the specimen. Because of the spatial separation, they will experience different optical path lengths as they pass through the specimen, which causes a phase change based on the thickness and refractive index of the specimen. Finally, the two beams are recombined into one. The combination of two beams leads to interference, due to the phase differences, giving bright-and-dark contrast signal in the images (Fig. 12B). Similar to PCM, DIC enables the visualization of transparent specimens without staining and has been widely used in biological imaging.

The resolution of an optical microscope is defined as the shortest distance between two objects that can be distinguished by the detector (Davidson & Abramowitz, 2002; Mertz, 2010). When the incident light passes through a specimen, it generates a diffraction pattern projected on the objective lens. This diffraction pattern appears as Airy disks, which are described by the point spread function (PSF): the greater the diffraction pattern, the more accurate the image. There are two parameters that determine the

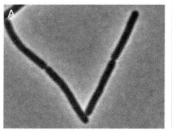
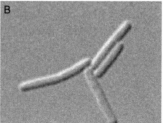

FIG. 12

Phase contrast images (A) and differential interference contrast image (B) of *B. subtilis* cells.

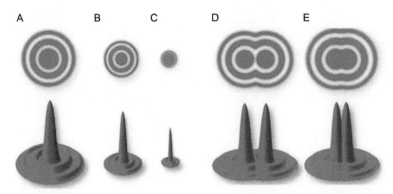

FIG. 13

2D (*upper*) and 3D (*bottom*) representation of Airy disks and resolution. (A–C) Size and related intensity profile of an Airy disk. The size of the observed diffraction pattern decreases from (A) to (C) when NA is increased or short-wavelength incident light is used. (D and E) Airy disks of two close objects that can (D) or cannot (E) be distinguished.

degree of the diffraction pattern: numerical aperture (NA) and incident light wavelength. The NA of a microscope determines the range of angles through which the objective can accept light. A higher NA value expands the range through which the diffraction pattern can be received, thus providing higher resolution. On the other hand, the wavelength of the incident light determines the size of the diffraction pattern (Fig. 13A–C). Given the same NA value, a greater level of the diffraction pattern can be received if its size is smaller. Therefore, an optical microscope using high NA objective and short-wavelength incident light provides high-resolution images for the specimens. The maximum resolution of a modern optical microscope can be about 200 nm. Microscope systems specialized for super-resolution images can even reach resolutions of about 50 nm. Examples include stimulated emission depletion microscopy and stochastic optical reconstruction microscopy (Hell & Wichmann, 1994; Rust, Bates, & Zhuang, 2006).

FM is a specialized optical microscopy technique for studying materials that generate fluorescence signal, either when they are coupled with fluorescent chemical or in its natural state (autofluorescence) (Davidson & Abramowitz, 2002; Mertz, 2010). To generate images, FM first irradiates the specimens with excitation light of a specific wavelength. Upon excitation, the specimens emit fluorescence light at longer wavelength. The emitted light signal then passes through an emission filter and is detected by a camera. The function of the emission filter is to filter out the reflected excitation light so that only the signal generated from the specimen will be visualized. FM has been applied to study organic materials, such as tissues and cells, as well as inorganic materials, such as semiconductors. With the combination of fluorescence labelling, FM facilitates the identification of cells and subcell components with a high degree of specificity amid the nonfluorescent material. If multiple fluorescence labelling is introduced to a given sample, it is even possible to study the interplay between the labelled materials, which allows researchers to investigate macromolecule interaction inside cells.

The chemical probes used for fluorescence labelling are known as fluorophores (or fluorochromes) (Terai & Nagano, 2013). When an incident light at specific wavelength is absorbed by a fluorophore, the energy level of the fluorophore is increased. The excited fluorophore will then generate fluorescence light to release the energy and get back to ground energy state. The emission light from the fluorophore has a longer wavelength compared to the excitation light because part of the excitation energy is released by a nonradiation pathway, such as intramolecular vibration and rotation. Because of the difference in wavelength, we can detect the emission light specifically by using optical filter sets. The growth of fluorophore development has greatly expanded the application of FM. There are hundreds of commercially available fluorophores with different colours and brightness for various biological targets. Example includes DAPI dye (4,6-diamidino-2-phenylindole) for DNA labelling (Kapuscinski, 1995) and FM™ dye for cell membrane labelling. The introduction of fluorescent protein fusion also significantly advanced biological studies because it enables the specific fluorescence labelling of proteins of interest, allowing in vivo investigation of proteins in cells and living organisms.

In order to study the PG structure and its biosynthesis, fluorescent molecules have been developed for PG-specific labelling. In the following sections, we will discuss the property and application of fluorescent probes for peptidoglycan-specific labelling. Such probes provide powerful approaches for studying PG. Based on their structural design, they have been applied to address various questions about PG structure and the biosynthesis.

3.2 FLUORESCENT MOLECULAR PROBES FOR PG-SPECIFIC LABELLING
3.2.1 Fluorescent antibiotics

Early efforts to develop a fluorescent probe for PG-specific labelling sought to take advantage of antibiotics that have a high binding affinity for PG. Vancomycin is a glycopeptide antibiotic that has been used as the last-line treatment against hospital infections caused by Gram-positive bacteria. Vancomycin binds to the D-Ala-D-Ala terminus of the stem peptide in nascent PG, resulting in a steric block of the transpeptidation reaction required for cross-linking (Hammes & Neuhaus, 1974). Although the D-Ala-D-Ala subunit in the stem peptide is present in nascent PG in almost all bacterial species (Fig. 14), in mature PG, the D-Ala-D-Ala terminus is modified either as a result of the cross-linking reaction or through the cleavage of the terminal D-Ala residue by carboxypeptidases. As a result, Daniel reasoned that vancomycin should bind to nascent PG, which has not been modified, with higher specificity than mature PG (Daniel & Errington, 2003). A fluorophore-coupled vancomycin can thus be used to probe the pattern of nascent PG insertion in bacteria. To test this idea, Daniel synthesized fluorescein-coupled vancomycin (Van-FL) and used it to label Gram-positive species. (This strategy is limited to Gram-positive organisms as the outer membrane of Gram-negative organisms blocks entry of

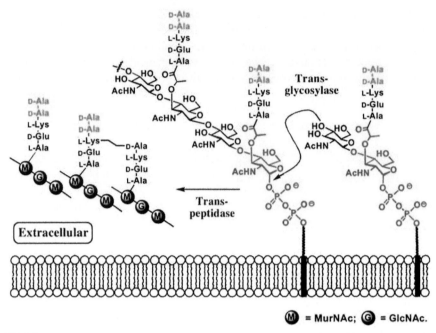

FIG. 14

Peptidoglycan binding sites of vancomycin (*red*) and ramoplanin (*blue*). *Left PG structure*, mature PG; *middle PG structure*, nascent PG; *right PG structure*, lipid II (Tiyanont et al., 2006).

vancomycin into the periplasm where PG synthesis is taking place.) Van-FL labelling in *B. subtilis* showed a diffused stripe pattern throughout the cells, indicative of circumferential insertion of new PG. The strongest signal was observed at the midcell indicative of an increased rate of PG synthesis and septum formation as the cell prepares to undergo division (Fig. 15A). The lack of signal at poles is in agreement with previous findings that *B. subtilis* undergoes little cell wall synthesis/turnover at its poles (Clarke-Sturman et al., 1989; de Pedro et al., 1997).

Streptococcus pneumoniae, however, has a very different labelling pattern from that of *B. subtilis*. It has been known that *S. pneumoniae* synthesizes PG specifically from equatorial rings that elongate and constrict simultaneously for cell elongation and division, respectively (Koch, 2000; Zapun, Vernet, & Pinho, 2008). Thus, Van-FL revealed a midcell labelling pattern in *S. pneumoniae* (Fig. 15B) (Daniel & Errington, 2003). In contrast, *S. coelicolor* is a polar-growing bacterium that forms nascent PG at cell poles. Microscopy images using Van-FL displayed a strong signal at the cell poles (Fig. 15C) (Daniel & Errington, 2003).

Since the intact D-Ala-D-Ala dipeptide is largely confined to nascent PG, labelling with Van-FL should be PG synthesis dependent. To confirm that Van-FL was labelling nascent PG, Van-FL labelling efficiency was evaluated under various conditions designed to impact the PG synthesis pathway. These studies revealed that the

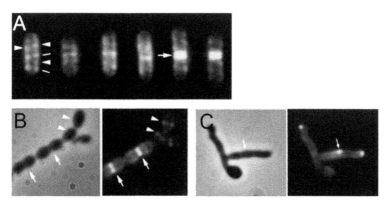

FIG. 15

Fluorescence microscopy images of bacterial species labelled with Van-FL. (A) Labelling pattern of *B. subtilis*. Cells in different division states were aligned from *left* (no division) to *right* (late division state). *White arrows* point out the observed striped signal in lateral and septal PG. (B) Labelling pattern of *S. pneumoniae*. *White arrows* indicate the labelling bands at the middle cell. (C) Labelling pattern of *S. coelicolor*. *White arrows* indicate subapical spots that present the nascent cell branch (Daniel & Errington, 2003).

depletion of proteins involved in PG synthesis (e.g. MurE) resulted in substantial reduction of the Van-FL labelling signal (Daniel & Errington, 2003). Also, treatment with antibiotics that inhibited PG precursor synthesis showed a clear suppression of Van-FL labelling.

Fluorescent ramoplanin derivatives, in conjunction with FM, have also been used to visualize PG synthesis. Although Van-FL has high specificity for the D-Ala-D-Ala stem peptide subunit in nascent PG, it is not possible to completely rule out Van-FL binding to mature PG (e.g. stem peptides that have not been cross-linked). Ramoplanin binds to the reducing end of the nascent glycan strand, which is only found at the initiation sites of lipid II insertion (Fig. 14) (Tiyanont et al., 2006). The labelling results using fluorescent ramoplanin in *B. subtilis* suggested a PG growth pattern consistent with the Van-FL data where diffused circumferential insertion of nascent PG was observed throughout the cells. The strongest fluorescence signal was observed at the division site, while the weakest signal was observed at the cell poles. Since fluorescently modified antibiotics also have toxic effects on bacteria and cell growth, careful optimization of labelling concentrations is required in order to prevent aberrant cell growth patterns (Tiyanont et al., 2006).

3.2.2 Fluorescent wheat germ agglutinin

Fluorescent wheat germ agglutinin (FWGA) is another tool that has been used for PG labelling. WGA is a protein found abundantly in wheat kernels. It has a high binding affinity for *N*-acetyl-glucosamine (GlcNAc), one of the essential components of PG glycan strands, and thus can be used to probe PG specifically. FWGA was first used as an alternative to the Gram stain and used to distinguish Gram-negative and Gram-positive bacteria (Sizemore, Caldwell, & Kendrick, 1990). FWGA can easily access the peptidoglycan layer in Gram-positive organisms, but it cannot penetrate the outer membrane in Gram-negative organisms. As a result, Gram-positive species can be efficiently labelled by FWGA, but Gram-negative species cannot. Because of its high binding affinity for N-acetylglucosamine subunits embedded in glycan strands, FWGA enables PG labelling in very low concentrations and short incubations time when compared to traditional Gram staining.

FWGA has also been employed to study the PG synthesis activity. A long-standing question regarding bacterial growth is how bacterial cells regulate PG synthesis while precisely maintaining cell morphology. MreB, a cytoskeletal protein, coordinates PG synthesis along the lateral wall by recruiting the requisite PG synthesis machinery (Soufo & Graumann, 2003; Typas et al., 2012). However, the precise means by which MreB regulates PG synthesis is not fully understood. In order to gain insight into this question, Ursell studied the turnover pattern of FWGA-labelled PG in *E. coli* using pulse-chase methods and time-lapse FM (Ursell et al., 2014). In this study, *E. coli* cells were first labelled with FWGA and then transferred to FWGA-free medium for an additional incubation. The PG turnover pattern was computationally analysed in order to map the growth, geometry, and cytoskeletal organization at subcellular resolution. The results revealed that nascent PG insertion occurs in a heterogeneous fashion and is correlated with MreB spatiotemporally.

3.3 DEVELOPMENT OF METABOLIC PROBES FOR PG-SPECIFIC LABELLING: FDAAs

Although the use of fluorescent antibiotics and FWGA has advanced our knowledge of PG synthesis dynamics, their application in studies of PG synthesis dynamics is restricted due to their labelling/signalling limitations. In the case of fluorescently modified antibiotics, the concentration of the probes needs to be carefully optimized to protect the cell from toxic effects caused by the antibiotics (Tiyanont et al., 2006). Care must also be taken in experiments utilizing these probes in order to rule out the possibility of erroneous PG synthesis patterns resulting from antibiotic inhibition. With respect to FWGA labelling, studies of PG synthesis dynamics were carried out by studying the reduction of WGA signal resulting from new PG insertion (Ursell et al., 2014). This signal reduction results in decreased resolution.

While these probes significantly advanced the ability to spatiotemporally visualize PG synthesis, it is clear that each method had limitations that precluded their widespread use. In an effort to identify a chemical probe to circumvent these limitations, the VanNieuwenhze and Brun laboratories designed and developed fluorescent D-amino acids (FDAAs). These small-molecule probes have found widespread use for covalently labelling PG in live bacterial cells. The discovery of these probes and their applications will be described in the sections that follow.

3.3.1 Development of D-amino acid-based fluorescent probes

Organisms in all kingdoms predominantly utilize L-amino acids for the synthesis of proteins and other biomolecules. Bacteria, however, also utilize D-amino acids and these amino acids (e.g. D-Glu, D-Ala) are typically found in PG. Furthermore, Lam reported the interesting observation that bacteria produce a variety of D-amino acids (Lam et al., 2009). These D-amino acids, although rare in nature, can accumulate in stationary phase culture supernatants at millimolar concentrations. *E. coli* can also incorporate exogenous D-cysteine in its PG. In addition, Kahne and Walker (Lupoli et al., 2011) reported that some D-amino acids, but not their L-isomers, can be incorporated specifically into PG fragments by *E. coli* PBP1a. Since bacteria can utilize D-amino acids for the construction of PG and various D-amino acids can be incorporated into PG via amino acid exchange reactions, we reasoned that it might be possible to use fluorescently modified D-amino acids as specific PG labelling probes.

A series of D-amino acid-based metabolic probes (FDAAs) (Kuru et al., 2012; Kuru, Tekkam, Hall, Brun, & VanNieuwenhze, 2015) were developed (Fig. 16) that specifically label PG in a variety of bacteria. These D-amino acid derivatives, although structurally diverse, are incorporated into PG at sites where active growth/remodelling is occurring. The labelling reactions utilizing FDAAs are specific, while the FDAAs themselves are nontoxic (Fig. 17) (Kuru et al., 2012).

In these experiments, bacterial cells were incubated with the corresponding FDAAs for 1 h. Following incubation, the cells were washed (to remove unincorporated probe) and fixed in preparation for visualization via FM. Sacculi were also purified and isolated from the labelled sample. The labelling patterns were consistent for both labelled cells and sacculi purified from the labelled population, demonstrating that fluorescence was specifically localized within PG structures (Fig. 17A). MS analysis of purified and digested sacculi further confirmed that FDAAs were incorporated into PG peptide

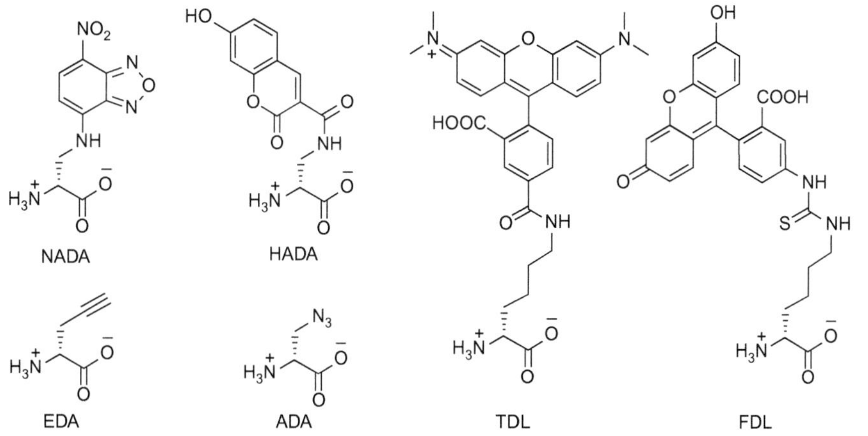

FIG. 16

D-Amino acid-based fluorescent probes.

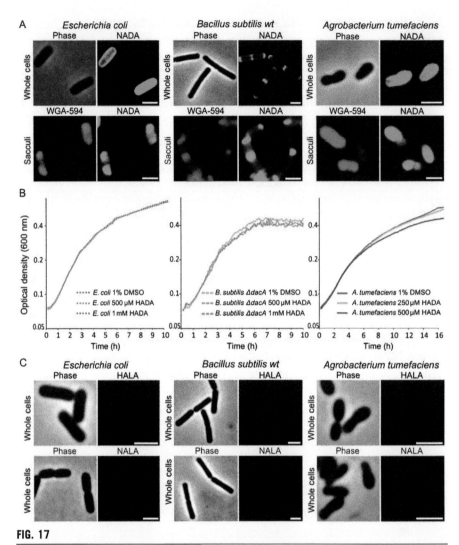

FIG. 17

Specific labelling of PG by FDAAs is nontoxic (Kuru et al., 2012).

stems. Since cytotoxicity can sometimes be an issue for metabolic probes, a growth curve experiment was performed (Fig. 17B). Bacterial cells were incubated with FDAAs at various concentrations while cell growth was monitored. FDAAs did not exhibit any significant toxicity when compared with the control group. In order to test the specificity of FDAA incorporation, a control labelling experiment was conducted utilizing the enantiomeric fluorescent L-amino acids (FLAAs). FLAAs do not label bacteria (Fig. 17C), indicating that the PG synthesis machinery is specific for FDAAs. In both Gram-positive and Gram-negative species, only FDAAs, and not FLAAs, were covalently incorporated into PG.

FDAAs were found to label a broad array of evolutionarily diverse organisms (Kuru et al., 2012). This method for PG labelling appeared to be universal and adaptable to a variety of model organisms (e.g. *E. coli* and *B. subtilis*) (Fig. 18).

FDAAs addressed problems encountered with other fluorescent PG labelling methods. When compared to fluorescent antibiotics (e.g.; Van-FL), FDAAs do not display significant cytotoxicity and thus would not be anticipated to interfere with the normal cellular metabolism of bacteria. As a result, the possibility of observing aberrant growth patterns due to inhibition of cellular processes is much lower. When compared with FWGA labelling, FDAAs are "gain of signal probes" instead of "loss of signal probes". As a result, they can be used to "chase" new PG synthesis patterns and the signal gain is indicative of, and in accordance with, new PG synthesis activity, whereas FWGA labelling provides a signal that fades with the new PG synthesis activity.

Another advantage of FDAA probes is their synthetic accessibility. The synthetic routes for representative FDAAs are illustrated in Fig. 19. The synthetic routes are modular and utilize readily available D-amino acid precursors. Each of the amino acid precursors has a side chain containing a nucleophile capable of capturing an activated fluorophore (often available from commercial sources). HADA and NADA utilize *N*-Boc-D-2,3-diaminopropionic acid (Boc-D-Dap-OH) as the backbone D-amino acid; their respective fluorogenic groups (7-hydroxycoumarin-3-carboxylic acid, HCC-OH, for HADA and 4-chloro-7-nitrobenzodurazan, NBD-Cl, for NADA) are attached to the side chain of Boc-D-Dap-OH. FDL and TDL are synthesized from Boc-D-Lys-OH with FITC (for FDL) or 5 (and 6-)-carboxytetramethylrhodamine succinimidyl ester (TAMRA-OSu, for TDL), respectively. These procedures can be easily and reproducibly executed in a basic organic chemistry laboratory. Moreover, the modular design of the synthetic routes can readily accommodate a variety of fluorophores or other functionalities (Kuru et al., 2015).

In a typical FDAA labelling protocol, early exponential phase cells are exposed to FDAAs for either 2-5% of doubling time (short pulse) or one to two generations (long pulse), followed by fixing with 70% ethanol and a thorough washing protocol to remove unincorporated FDAA probe (washing $3 \times$ with PBS). The fixed and washed cells are resuspended in PBS and then subjected to fluorescent microscopy. For clickable probes like EDA and ADA (Fig. 16), an additional click reaction step and a washing step need to be amended after regular fix–wash cycle. A complete description of FDAA synthesis and labelling protocols has been published (Kuru et al., 2015).

3.3.2 Proposed mechanism of FDAA labelling

Since it has been demonstrated that FDAAs are incorporated into PG in a growth-dependent fashion, recent efforts have focused on elucidating the mechanism(s) for their incorporation (Kuru et al., 2012). Since available evidence suggests that FDAAs are incorporated metabolically, the question arises as to which PG synthesis pathway is used, or which PG synthesis protein(s) is/are responsible for FDAA incorporation?

28 **CHAPTER 1** Methods for visualization of peptidoglycan biosynthesis

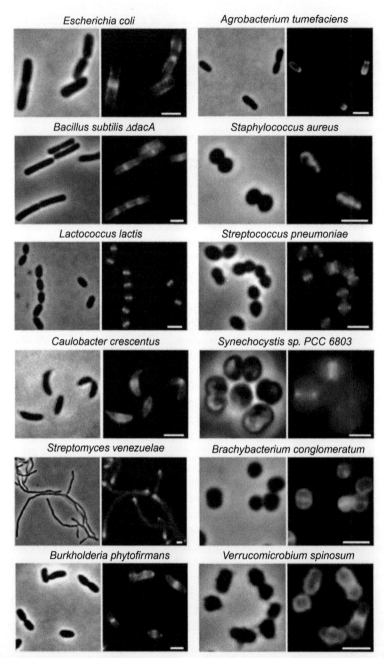

FIG. 18

Short pulse labelling by HADA in various bacteria. Scale bars: 2 mm (Kuru et al., 2012).

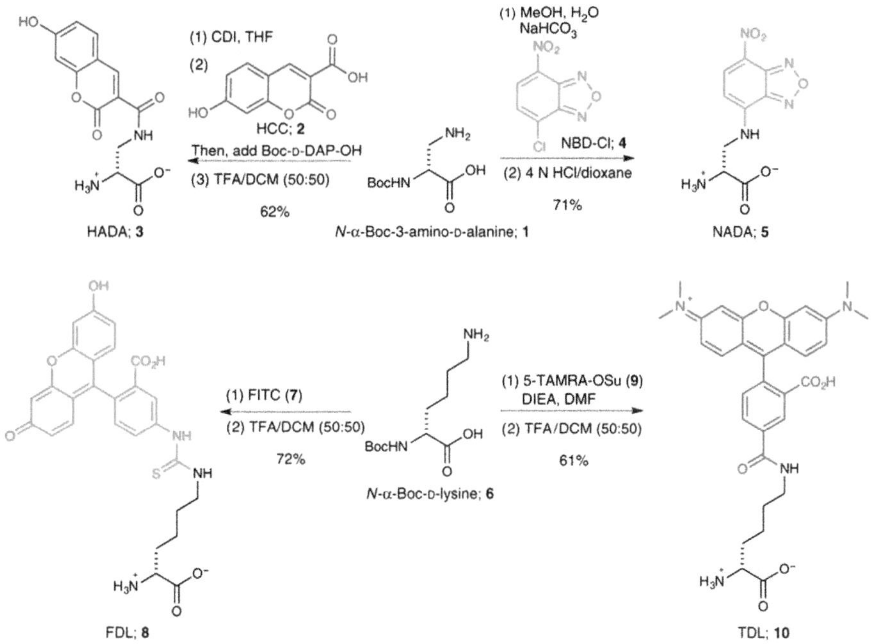

FIG. 19

Modular syntheses of FDAAs. *DMF*, dimethylformamide; *THF*, tetrahydrofuran (Kuru et al., 2015).

Cells pretreated with sublethal dosage of β-lactam antibiotics do not retain any fluorescent signal after FDAA labelling, indicating that the labelling could be dependent on transpeptidation. Based on this information, as well as other experimental data gathered in our laboratory, there are two possible pathways for FDAA probes to get incorporated in bacteria (Fig. 20): (1) incorporation mediated by the PBP D,D-transpeptidases (Sauvage et al., 2008); or (2) incorporation through the cooperative activity of PBP D,D-carboxypeptidases (Sauvage et al., 2008) and L,D-transpeptidases (Ldts) (De & McIntosh, 2012; Magnet, Dubost, Marie, Arthur, & Gutmann, 2008; Sanders & Pavelka, 2013; van Heijenoort, 2011).

In general, PBPs (Sauvage et al., 2008) can be divided into two main categories: the high-molecular weight (HMW) PBPs and the low-molecular weight (LMW) PBPs. HMW PBPs are PG synthases that are responsible for the polymerization and cross-linking of PG, and are usually multimodular proteins. LMW PBPs, on the other hand, are PG hydrolases, with the exception of PBP4 in *S. aureus*. The LMW PBPs are responsible for the remodelling of PG (van Heijenoort, 2011).

Within the category of HMW PBPs, there are two classes of PBPs based on their structure and enzymatic properties. Class A HMW PBPs are bifunctional PBPs. They have a penicillin-binding (PB) domain (Yeats, Finn, & Bateman, 2002; Zapun, Contreras-Martel, & Vernet, 2008) near their C-terminus that has

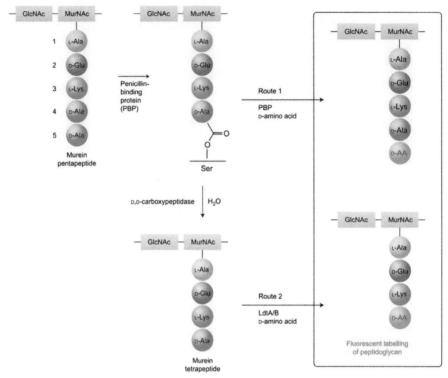

FIG. 20

Potential pathways for FDAA incorporation (Bugg, 2012).

D,D-transpeptidation activity and a glycosyltransferase domain in their N-terminus. In the case of class B HMW PBPs, the C-termini are similar, while the N-terminal domains often contain recognition modules that play important roles in cell morphology (Sauvage et al., 2008).

LMW PBPs are often referred to as class C PBPs. They also have a PB domain, but instead of D,D-transpeptidase activity, they are frequently D,D-carboxypeptidases and/or D,D-endopeptidases. Interestingly, PBP4 in *S. aureus*, though a class C PBP, has both D,D-transpeptidase and D,D-carboxypeptidase activities (Qiao et al., 2014).

The PB domains in all PBPs are of the active-site serine penicillin-recognition enzyme family (Yeats et al., 2002; Zapun, Contreras-Martel, & Vernet, 2008). Without exception, they all have an active-site serine as a nucleophile. The mechanism of transpeptidation is shown in Fig. 21 (Oliva, Dideberg, & Field, 2003). The active-site serine first attacks PG at the amide bond between D-Ala$_4$ and D-Ala$_5$ and forms an acyl-enzyme complex, releasing D-Ala$_5$ in the process. Following this, an adjacent stem peptide, with a free amino group in the side chain of the third amino acid (or a free amino group on the terminal amino acid of an attached bridge link), acts as an acyl acceptor and carries out a nucleophilic attack on the intermediate acyl-enzyme complex. The resulting product is cross-linked PG with a 4–3

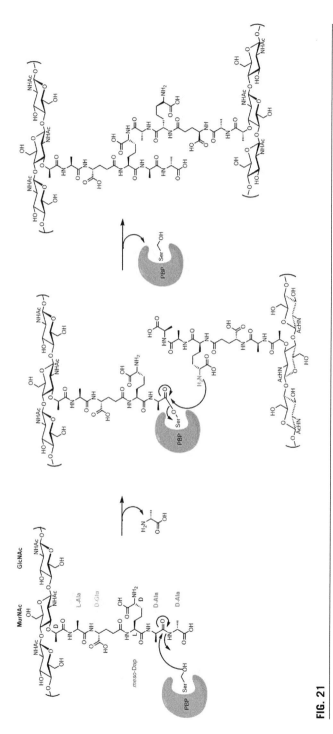

FIG. 21

Mechanisms of penicillin-binding protein-mediated transpeptidation.

Table 2 List of PBPs in *E. coli*

Name	Class	Enzymatic Function	Role
PBP1a	A	Bifunctional GTase and TPase	Elongation
PBP1b	A	Bifunctional GTase and TPase	Division
PBP1c	A	GTase only	Unknown
PBP2	B	TPase	Elongation
PBP3	B	TPase	Division
PBP4	C	CPase and EPase	
PBP4b	C	CPase	
PBP5	C	CPase	Major PBP, control cell shape
PBP6	C	CPase	Stabilize PG, high in stationary phase
PBP6b	C	CPase	
PBP7	C	EPase	Major PBP
AmpH	C	CPase and EPase	Control cell shape

linkage (D-Ala$_4$ from one strand and AA$_3$ (*meso*-DAP, Lys, etc.) from the adjacent strand). The carboxypeptidases/endopeptidases follow a similar mechanism; the only difference is that instead of the capture of the acyl-enzyme intermediate by another stem peptide acceptor, H$_2$O will be the acyl acceptor to attack the acyl-enzyme complex, resulting in hydrolysis. D,D-Carboxypeptidases cleave D-Ala$_5$, leaving a tetrapeptide PG strand with D-Ala$_4$ at the end, while D,D-endopeptidases cleave the amide bond between D-Ala$_4$ from one peptide stem and sidechain D-center of *meso*-DAP$_3$ from the other peptide stem (Table 2).

The PG synthesis machinery also has another class of enzymes, although much less studied, that may also facilitate FDAA incorporation. Mainardi et al. (2005) discovered an enzyme termed Ldt$_{fm}$, in a β-lactam-resistant mutant of *Enterococcus faecium*. This enzyme can catalyse an in vitro reaction that results in the introduction of a cross-link between L-Lys$_3$ and D-iAsn-L-Lys$_3$. This novel linkage may account for the β-lactam resistance, as the tetrapeptide substrate required for this reaction is distinct from the pentapeptide substrate for PBP-catalysed reactions. Further analysis of active-site residue, and mutation studies, found that this enzyme utilizes an active-site cysteine residue to perform a transpeptidation reaction similar to those mediated by the PBPs. Upon sequence comparison, Ldt$_{fm}$ homologs are also found in bacteria that are genetically and taxonomically distant from *E. faecium*, which indicate that this mode of cross-linking could be associated with resistance development in other bacteria, especially pathogenic species.

Ldts have two functions in Gram-negative bacteria: cross-linking of adjacent peptide strands or anchoring Braun lipoproteins onto PG (Magnet et al., 2007, 2008). The cross-linking reaction catalysed by Ldts is initiated by recognition of tetrapeptide substrate obtained after loss of the terminal D-ala residue through the action of a D,D-carboxypeptidase. The active-site cysteine then attacks the amide bond between *meso*-DAP$_3$ and D-Ala$_4$, followed by release of D-Ala$_4$, to form a PG-Ldt thioester intermediate. This acyl-enzyme (thioester) intermediate is then captured by an acyl acceptor, the ε-amino group from *meso*-DAP$_3$ on an adjacent PG peptide stem, resulting in the introduction of a 3,3-cross-link. In *E. coli*, Ldts are believed to be responsible for approximatetly 2–5% of all cross-links (Magnet et al., 2008).

In a representative Gram-negative organism, *E. coli*, it is estimated that only 20–30% of the stem peptide is cross-linked (Magnet et al., 2008). These cross-links are of two distinct types that are catalysed by two kinds of enzymes. The most common 4–3 linkage (Fig. 22A) is formed by PBPs (classes A and B) and the 3–3 linkage (Fig. 22C), which cross-links L-center of the third position *meso*-DAP from one stem to the D-center of the third position *meso*-DAP in another stem, is the product of LD-transpeptidases (Ldts) (Table 3).

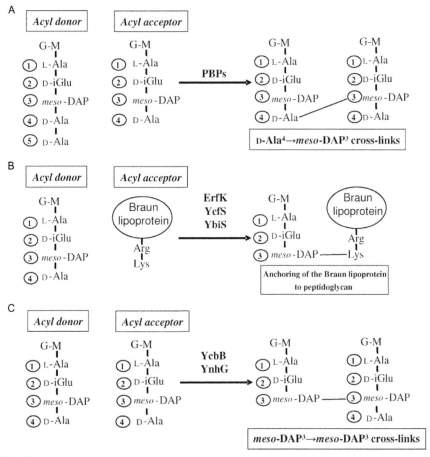

FIG. 22

Comparison of PBP and Ldt cross-links (Magnet et al., 2008). (A) 4–3 cross-links formed by Ddt PBPs; (B) Braun lipoprotein anchoring by Ldt ErfK, YcfS, and YbiS; (C) 3–3 cross-links formed by Ldt YcbB and YnhG.

Table 3 List of Ldts in *E. coli*

Name	Enzyme	Function	Notes
ErfK/LdtA	Ldt1	Braun-lipoprotein ligation	
YbiS/LdtB	Ldt2	Braun-lipoprotein ligation	Major enzyme for lipoprotein ligation
YcfS/LdtC	Ldt4	Braun-lipoprotein ligation	
YcbB/LdtD	Ldt3	3–3 linkage	Major enzyme for 3–3 linkage
YnhG/LdtE	Ldt5	3–3 linkage	
YafK	Ldt6	Unknown	Involved in the biofilm formation in enteroaggregative *E. coli*

The originally proposed mechanism for FDAA probe incorporation is that they function as acyl acceptors in PBP-mediated PG cross-linking reactions and that they are incorporated at the terminal (fifth position) of the stem peptide of nascent PG. Ldt-mediated incorporation resulted in FDAA incorporation at the fourth position of tetrapeptide stems. As suggested by the HPLC-MS/MS results of purified and digested sacculi, FDAAs are exclusively incorporated in the fourth position in *E. coli* and *Agrobacterium tumefaciens*, which suggest that Ldts play a major role in FDAA incorporation in Gram-negtive organisms (Kuru et al., 2012). PBPs also play important roles in the FDAA incorporation, particularly in Gram-positive organisms. For example, sacculi digestion experiments, followed by tandem MS analysis, revealed that Gram-positive organisms (e.g. *B. subtilis*) incorporated FDAAs at the fifth position of the stem peptide. Furthermore, *dacA* deletion strains of *B. subtilis* exhibit much stronger signals than wild-type strains, suggesting that the presence of DacA (a D,D-carboxypeptidase) may attenuate preexisting fluorescence incorporation. Further supporting this notion, a pronounced decrease in fluorescence incorporation is observed upon treatment with sublethal doses of β-lactam antibiotics.

4 APPLICATIONS OF FDAAs AND THEIR DERIVATIVES
4.1 TOOLS FOR STUDIES OF PG SYNTHESIS PATTERN AND DYNAMICS OF PG

One of the most powerful applications for FDAAs is their use in studying the spatial and temporal dynamics of PG synthesis (Kuru et al., 2012). Through implementation of various labelling techniques, FDAAs can reveal PG synthesis patterns in different bacteria. Fig. 23A shows a classical pulse-chase time-lapse experiment which can distinguish newly synthesized PG from parent cell PG. Here, *E. coli* and *B. subtilis* Δ*dacA* cells were first labelled with HADA (7-hydroxycoumarin-3-carboxylic acid-3-amino-D-alanine) for 1 h and then monitored by FM for the insertion of a new cell wall material. The labelled portion would be old PG and the insertion of new PG would be manifested by a decrease in the fluorescence signal. From the results, one can see that both *E. coli* and *B. subtilis* ΔdacA strains exhibit

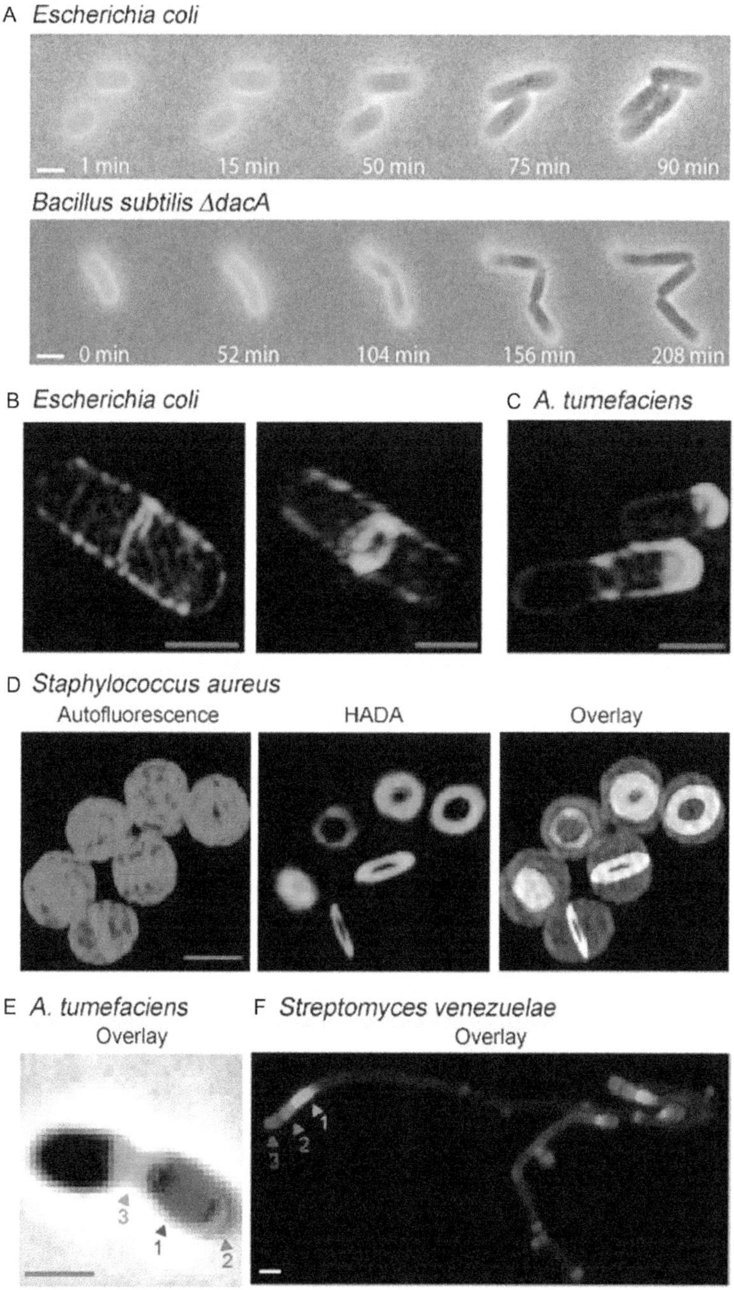

4 Applications of FDAAs and their derivatives

A *Escherichia coli*

Bacillus subtilis ΔdacA

B *Escherichia coli* C *A. tumefaciens*

D *Staphylococcus aureus*
Autofluorescence   HADA   Overlay

E *A. tumefaciens* F *Streptomyces venezuelae*
Overlay   Overlay

FIG. 23

FDAA labelling enables visualization of cell wall growth patterns in different bacteria (Kuru et al., 2012). *White scale bars*, 2 mm; *red scale bars*, 1 mm.

new PG insertion at the midcell. The *B. subtilis* ΔdacA strain also showed a strong signal at the cell poles. Super-resolution microscopy with short-pulsed cells enables us to directly observe newly synthesized PG with further enhanced spatial resolution (Fig. 23B–D). In *E. coli*, the septal region was strongly labelled by HADA, while in *A. tumefaciens*, a strong signal was observed at the cell poles, indicating active PG synthesis at these locations. With ovococcal cells like *S. aureus*, short pulses revealed various stages of constriction of the septal ring (Fig. 23D). Virtual time-lapse microscopy is another technique that can be facilitated with FDAA labelling. For example, bacterial cells can be labelled with an FDAA to provide a background signature of PG synthesis, then washed, and treated with a second FDAA (different emission wavelengths) for a second round of labelling. This process can be repeated with multiple FDAAs in subsequent rounds of labelling. As can be observed in labelling experiments conducted with *A. tumefaciens* and *Streptomyces venezuelae*, FDAAs can be used in virtual time-lapse microscopy applications to provide a chronological time stamp of PG synthesis (Fig. 23E and F).

4.2 PG RECYCLING/REMODELLING IN *STREPTOCOCCUS PNEUMONIAE*

FDAAs can also be used to study PG turnover (Boersma et al., 2015). PG is a dynamic structure that is constantly undergoing the dual processes of new PG synthesis and old PG degradation. During their life cycles, many bacteria, and in particular, Gram-negative bacteria, will turn over and possibly recycle fragments of old PG (Johnson, Fisher, & Mobashery, 2013). This is an important evolutionary survival advantage for bacteria since the fragments can induce antibiotic resistance gene expression (Fisher & Mobashery, 2014), can serve as signal molecules in the spore germination pathways (Bertsche, Mayer, Gotz, & Gust, 2015), and can also trigger innate immune response in the host (Boudreau, Fisher, & Mobashery, 2012). In order to avoid the latter, some bacteria will recycle free PG fragments in order to keep their abundance to a minimum.

In a study of *S. pneumoniae*, FDAAs were used to monitor PG dynamics, as well as morphological defects in several *S. pneumoniae* strains. By applying long-pulse-chase-new-labelling experiments, new and old PG was differentiated with two different FDAAs. *S. pneumoniae* IU1945 and IU3900 strains were labelled using the virtual time-lapse method (described earlier), first with HADA (pseudocoloured green, representing old PG) for three generations and then followed by labelling with NADA (pseudocoloured red, representing new PG) (Fig. 24). Live cells were then sampled for fluorescence localization at different time points after NADA was added. The results were consistent with minimum PG turnover: the old PG remains intensively labelled green (as hemisphere caps) and persisted into stationary phase. In addition, there was minimal, if any, overlap between old PG signals and new PG signals and the new PG signals seem to originate from septal regions where active PG synthesis is taking place.

Single-cell time-lapse microscopy was also applied to confirm the persistent hemispherical signals (Fig. 25). *S. pneumoniae* IU1945 was grown in BHI broth and then

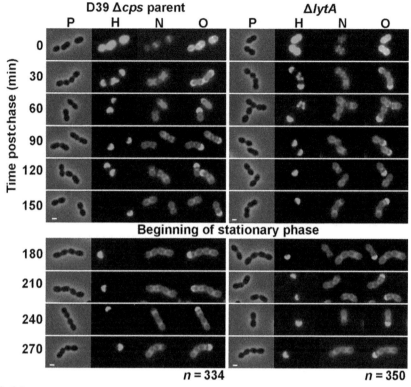

FIG. 24

Persistence of hemispheres of stable old PG in exponentially growing and early-stationary-phase cells of *S. pneumoniae* detected by FDAA long-pulse-chase-new-labelling experiments (Boersma et al., 2015). Scale bar = 1 μm. H, HADA labelling (old PG); N, NADA labelling (new PG synthesis); O, overlay of H and N images; P, phase-contrast image.

labelled with TADA (pseudocoloured orange) for about four generations. After washing, cells were resuspended in BHI (lacking FDAA) and then monitored via FM. The labelled four hemispheres in each image correspond to the original four hemispheres of the initial parent cell and the intensities of this hemispherical fluorescence did not seem to diminish over time, indicating minimum PG-remodelling activity.

FDAAs were also used to show a morphologically aberrant growth in PG-remodelling hydrolase mutants (Fig. 26). *S. pneumoniae* mutant strains with various hydrolase deletions or depletions were grown exponentially in BHI broth, labelled with HADA (pseudocoloured green; old PG), washed, and chased in the presence of NADA (pseudocoloured red; newly synthesized PG). Live cells were imaged by epifluorescence phase-contrast microscopy at the indicated time points after the removal of HADA (chase) and the addition of NADA.

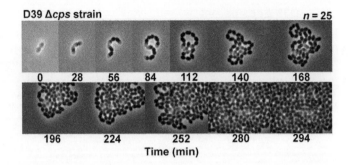

FIG. 25
Retention of hemispheres of stable TADA-labelled PG in single pneumococcal cells viewed by time-lapse microscopy (Boersma et al., 2015).

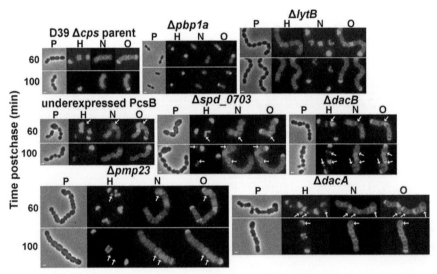

FIG. 26
Mutants deficient in PG hydrolases retain hemispheres of stable PG and show regions of inert PG at aberrant division points (Boersma et al., 2015). Scale bar = 1 μm. *H*, HADA labelling (old PG); *N*, NADA labelling (new PG synthesis); *O*, overlay of H and N images; *P*, phase-contrast image. *Arrows* indicate regions in kinked chains bounded by *green spots* of inert peptidoglycan, by *red spots* or *regions* of newly synthesized PG trapped between regions of inert green PG, or by *closely spaced red regions* of newly synthesized PG.

4.3 FDAAs REVEAL PG IN STRAINS NOT PREVIOUSLY KNOWN TO POSSESS IT

There are several bacterial strains that may possess PG, but for which the currently available technology to detect or analyse PG has been unable to verify or rule out its existence. *Chlamydia trachomatis* is one such organism.

It has long been known that *Chlamydia* is susceptible to β-lactam antibiotics (McCoy & Maurelli, 2006). Although PG has never been successfully detected nor purified in *Chlamydia*, genetic analysis and antibiotic effectiveness strongly implicate its presence. D-Cycloserine (Moulder, Novosel, & Officer, 1963) and penicillins (Tamura & Manire, 1968), whose mechanisms of action are directed at targets in the PG synthesis pathway, are effective against species in Chlamydiae phylum. Moreover, genetic analysis (Stephens et al., 1998) of *C. trachomatis*, *Chlamydophila pneumoniae*, and *Protochlamydia amoebophila* confirms that they posess most of the genes for PG biosynthesis, including PBPs. These conflicting findings have resulted in half-century long debate termed the "Chlamydia anomaly" (Moulder, 1993).

Although FDAAs were successful in labelling *Protochlamydia* (Pilhofer et al., 2013), when they were first applied to the labelling of *C. trachomatis*, they did not provide any evidence of labelled PG. In the labelling experiments with *C. trachomatis* and *Shigella flexneri*, EDA labelling did not result in fluorescent signals when attempts were made to capture the alkyne functional handle with a complementary azide-modified fluorescent probe via a click reaction. This failure was attributed to cleavage of the EDA residue, either by a carboxypeptidase or by a PBP during cross-linking, by the *C. trachomatis* PG synthesis machinery.

Another possible pathway for D-amino acid incorporation is the MurF pathway (Fig. 27). MurF (Beeby, Gumbart, Roux, & Jensen, 2013; Kawai et al., 2011; Vollmer et al., 2008; Weidenmaier & Peschel, 2008) is an ATP-dependent ligase, which couples the D-Ala-D-Ala dipeptide to the UDP-MurNAc-tripeptide substrate that ultimately provides the Park nucleotide. Considering the possibility that *C. trachomatis* can take up D-Ala-D-Ala dipeptide, dipeptide derivatives of FDAAs were then synthesized and investigated for their ability to label PG (Fig. 28). Due to their structural similarity to the D-Ala-D-Ala dipeptide, it was hypothesized that

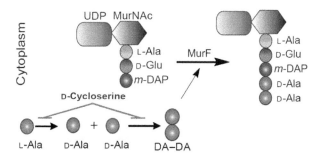

FIG. 27

MurF ligase mode of action. MurF is UDP-N-acetylmuramoyl-tripeptide-D-alanyl-D-alanine ligase. It performs the ligation of D-Ala-D-Ala dipeptide and UDP-MurNAc-tripeptide. The source of D-Ala-D-Ala is also shown here. L-Alanine in cytoplasm is converted to D-Ala by alanine racemase (Alr), and then two D-ala are coupled by D-Ala-D-Ala ligase (Ddl). Both Alr and Ddl can be inhibited by D-cycloserine (Liechti et al., 2014).

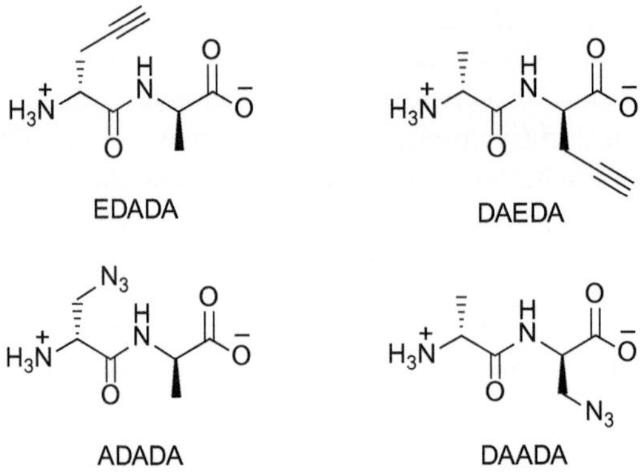

FIG. 28

Structures of dipeptide derivatives of FDAAs.

EDADA and DAEDA may be able to hijack MurF for incorporation into PG during the cytoplasmic stages of PG synthesis.

Cellular uptake of the dipeptide probes was examined by a plaque assay (Liechti et al., 2014). The design of this assay is to test whether an exogenous compound can support the growth of certain auxotroph. By adding D-cycloserine (DCS) in the media, bacteria can behave as D-Ala-D-Ala auxotroph because D-cycloserine effectively halts the pathway in which endogenous D-Ala-D-Ala is made. Addition of an exogenous D-Ala-D-Ala dipeptide derivative can, in principle, suppress DCS-initiated growth inhibition. These experiments revealed that Chlamydial cell growth could be rescued by both EDADA and DAEDA during treatment with D-cycloserine.

Despite being able to rescue growth in the plague assay, DAEDA was unable to label *Chlamydia*, a result initially attributed to a high rate of PG processing. As a result, EDADA was then investigated for its ability to label *C. trachomatis* within its host cell. Mouse fibroblast L2 cells were infected with *C. trachomatis* for 18 h in the presence of EDADA, and the subsequent click chemistry step revealed the localization of the incorporated probes. As shown in Fig. 29, *C. trachomatis* cells were labelled; specifically, the metabolically active reticulate body (RB) within host. The labelling revealed a ring-like shape around an apparent cellular division plane. This was the first observation and identification of the Chlamydial peptidoglycan structure, and it puts an end to a half-century debate over the existence of peptidoglycan in pathogenic *Chlamydia*. To further examine the nature of the labelling target, PG synthesis inhibitors were added during the labelling protocol. In the presence of ampicillin, the labelling was severely compromised, thus suggesting a labelling mechanism that relied upon a PBP-mediated reaction.

Using a similar approach, PG was also detected for the first time in an anaerobic ammonium-oxidizing (anammox) planctomycete, *Kuenenia stuttgartiensis* (van Teeseling et al., 2015).

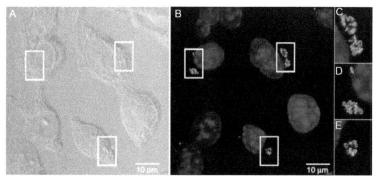

FIG. 29

Chlamydia peptidoglycan labelling by dipeptide probes. (A) Differential interference contrast (DIC) and fluorescent (B–E) microscopy of L2 cells infected for 18 h with *C. trachomatis* in the presence of the dipeptide probe EDA-DA (1 mM). Subsequent binding of the probe to an azide-modified Alexa Fluor 488 (green) was achieved via click chemistry (Liechti et al., 2014).

CONCLUSIONS

In this chapter, we have reviewed the current methods for PG visualization, including labelling with our new FDAA probes. Nonfluorescent methods like EM and AFM have inherent advantages in resolution, but they also have several limitations. Most SEMs and TEMs can only generate 2D image with greyscale colouring, though pseudo-3D imaging can allow for partially constructing "sideviews". Also, most EMs require vacuum for observation, which make it impossible to observe live samples. When compared to EM, AFM can generate 3D images with live samples, but its scanning rate limited its application for studying the highly dynamic PG structure. Fluorescent methods like fluorescent antibiotics and FWGA are great tools for studying PG dynamics, though fluorescent antibiotics like Van-FL are toxic to cells and PG dynamics may be altered as a consequence of such treatment. FWGA-labelling results are hard to interpret since the nascent PG insertion is presented as a reduction of FWGA signal.

FDAAs are metabolic probes that are designed for labelling the PG of live bacteria. Their incorporation pathway, though not fully determined, is PG specific. As promising as these probes have proven for elucidating the spatial and temporal dynamics of PG synthesis, we also anticipate the continued expansion of the utility of this class of chemical probes. First and foremost, because of the modular nature of FDAA synthesis, other fluorogenic groups can be installed on the sidechain to complete a full colour pallet that will enable multiple novel applications. Second, since FDAAs are incorporated metabolically, mechanistic probes that can be turned on only upon their incorporation into PG can be envisioned. These mechanistic probes could bypass the requirement for fixing and washing and could enable the study of

PG dynamics in live bacteria and in real time. This is currently an active area of research and strategies towards this end will be disclosed in the near future.

In conclusion, we hope that this chapter has convinced you of the utility of FDAAs in the study of peptidoglycan synthesis and dynamics. These are a remarkable set of tools that are capable of providing heretofore-unavailable insights into PG synthesis and dynamics. The widespread use and adoption of these tools provide further evidence of their novelty and utility.

REFERENCES

Allison, D. P., Mortensen, N. P., Sullivan, C. J., & Doktycz, M. J. (2010). Atomic force microscopy of biological samples. *Wiley Interdisciplinary Reviews. Nanomedicine and Nanobiotechnology, 2*(6), 618–634. http://dx.doi.org/10.1002/wnan.104.

Bartesaghi, A., Merk, A., Banerjee, S., Matthies, D., Wu, X., Milne, J. L., & Subramaniam, S. (2015). 2.2 Å resolution cryo-EM structure of beta-galactosidase in complex with a cell-permeant inhibitor. *Science, 348*(6239), 1147–1151. http://dx.doi.org/10.1126/science.aab1576.

Baum, E. Z., Crespo-Carbone, S. M., Foleno, B. D., Simon, L. D., Guillemont, J., Macielag, M., & Bush, K. (2009). MurF inhibitors with antibacterial activity: Effect on muropeptide levels. *Antimicrobial Agents and Chemotherapy, 53*(8), 3240–3247. http://dx.doi.org/10.1128/aac.00166-09.

Beeby, M., Gumbart, J. C., Roux, B., & Jensen, G. J. (2013). Architecture and assembly of the Gram-positive cell wall. *Molecular Microbiology, 88*(4), 664–672. http://dx.doi.org/10.1111/mmi.12203.

Bertsche, U., Mayer, C., Gotz, F., & Gust, A. A. (2015). Peptidoglycan perception—Sensing bacteria by their common envelope structure. *International Journal of Medical Microbiology, 305*(2), 217–223. http://dx.doi.org/10.1016/j.ijmm.2014.12.019.

Binnig, G., Rohrer, H., Gerber, C., & Weibel, E. (1982). Surface studies by scanning tunneling microscopy. *Physical Review Letters, 49*(1), 57–61. Retrieved from http://link.aps.org/doi/10.1103/PhysRevLett.49.57.

Blackman, S. A., Smith, T. J., & Foster, S. J. (1998). The role of autolysins during vegetative growth of Bacillus subtilis 168. *Microbiology, 144*(Pt 1), 73–82. http://dx.doi.org/10.1099/00221287-144-1-73.

Boersma, M. J., Kuru, E., Rittichier, J. T., VanNieuwenhze, M. S., Brun, Y. V., & Winkler, M. E. (2015). Minimal peptidoglycan (PG) turnover in wild-type and PG hydrolase and cell division mutants of Streptococcus pneumoniae D39 growing planktonically and in host-relevant biofilms. *Journal of Bacteriology, 197*(21), 3472–3485. http://dx.doi.org/10.1128/JB.00541-15.

Boneca, I. G., Huang, Z. H., Gage, D. A., & Tomasz, A. (2000). Characterization of Staphylococcus aureus cell wall glycan strands, evidence for a new beta-N-acetylglucosaminidase activity. *Journal of Biological Chemistry, 275*(14), 9910–9918.

Boudreau, M. A., Fisher, J. F., & Mobashery, S. (2012). Messenger functions of the bacterial cell wall-derived muropeptides. *Biochemistry, 51*(14), 2974–2990. http://dx.doi.org/10.1021/bi300174x.

Bouhss, A., Crouvoisier, M., Blanot, D., & Mengin-Lecreulx, D. (2004). Purification and characterization of the bacterial MraY translocase catalyzing the first membrane step of peptidoglycan biosynthesis. *Journal of Biological Chemistry, 279*(29), 29974–29980. http://dx.doi.org/10.1074/jbc.M314165200.

Bratkovic, T., Lunder, M., Urleb, U., & Strukelj, B. (2008). Peptide inhibitors of MurD and MurE, essential enzymes of bacterial cell wall biosynthesis. *Journal of Basic Microbiology, 48*(3), 202–206. http://dx.doi.org/10.1002/jobm.200700133.

Bugg, T. D. H. (2012). Biosynthesis: Imaging cell-wall biosynthesis live. *Nature Chemistry, 5*(1), 10–12. http://dx.doi.org/10.1038/nchem.1538.

Cetinkaya, Y., Falk, P., & Mayhall, C. G. (2000). Vancomycin-resistant enterococci. *Clinical Microbiology Reviews, 13*(4), 686–707. Retrieved from http://www.ncbi.nlm.nih.gov/pmc/articles/PMC88957/.

Chao, Y., & Zhang, T. (2011). Optimization of fixation methods for observation of bacterial cell morphology and surface ultrastructures by atomic force microscopy. *Applied Microbiology and Biotechnology, 92*(2), 381–392. http://dx.doi.org/10.1007/s00253-011-3551-5.

Clarke-Sturman, A. J., Archibald, A. R., Hancock, I. C., Harwood, C. R., Merad, T., & Hobot, J. A. (1989). Cell wall assembly in Bacillus subtilis: Partial conservation of polar wall material and the effect of growth conditions on the pattern of incorporation of new material at the polar caps. *Journal of General Microbiology, 135*(3), 657–665. http://dx.doi.org/10.1099/00221287-135-3-657.

Daniel, R. A., & Errington, J. (2003). Control of cell morphogenesis in bacteria: Two distinct ways to make a rod-shaped cell. *Cell, 113*(6), 767–776.

Davidson, M. W., & Abramowitz, M. (2002). *Optical microscopy. Encyclopedia of imaging science and technology*. John Wiley & Sons, Inc.

de Kruijff, B., van Dam, V., & Breukink, E. (2008). Lipid II: A central component in bacterial cell wall synthesis and a target for antibiotics. *Prostaglandins, Leukotrienes, and Essential Fatty Acids, 79*(3–5), 117–121. http://dx.doi.org/10.1016/j.plefa.2008.09.020.

de Pedro, M. A., Quintela, J. C., Holtje, J. V., & Schwarz, H. (1997). Murein segregation in Escherichia coli. *Journal of Bacteriology, 179*(9), 2823–2834.

De Petris, S. (1967). Ultrastructure of the cell wall of Escherichia coli and chemical nature of its constituent layers. *Journal of Ultrastructure Research, 19*(1), 45–83.

De, S., & McIntosh, L. P. (2012). Putting a stop to L,D-transpeptidases. *Structure, 20*(5), 753–754. http://dx.doi.org/10.1016/j.str.2012.04.005.

Dmitriev, B. A., Toukach, F. V., Schaper, K. J., Holst, O., Rietschel, E. T., & Ehlers, S. (2003). Tertiary structure of bacterial murein: The scaffold model. *Journal of Bacteriology, 185*(11), 3458–3468.

Duncan, K., van Heijenoort, J., & Walsh, C. T. (1990). Purification and characterization of the D-alanyl-D-alanine-adding enzyme from Escherichia coli. *Biochemistry, 29*(9), 2379–2386.

Fisher, J. F., & Mobashery, S. (2014). The sentinel role of peptidoglycan recycling in the beta-lactam resistance of the Gram-negative Enterobacteriaceae and Pseudomonas aeruginosa. *Bioorganic Chemistry, 56*, 41–48. http://dx.doi.org/10.1016/j.bioorg.2014.05.011.

Formanek, H., & Formanek, S. (1970). Specific staining for electron microscopy of murein sacculi of bacterial cell walls. *European Journal of Biochemistry, 17*(1), 78–84.

Gan, L., Chen, S., & Jensen, G. J. (2008). Molecular organization of Gram-negative peptidoglycan. *Proceedings of the National Academy of Sciences of the United States of America, 105*(48), 18953–18957. http://dx.doi.org/10.1073/pnas.0808035105.

Ghosh, A. S., Chowdhury, C., & Nelson, D. E. (2008). Physiological functions of D-alanine carboxypeptidases in Escherichia coli. *Trends in Microbiology, 16*(7), 309–317. http://dx.doi.org/10.1016/j.tim.2008.04.006.

Hammes, W. P., & Neuhaus, F. C. (1974). On the mechanism of action of vancomycin: Inhibition of peptidoglycan synthesis in Gaffkya homari. *Antimicrobial Agents and Chemotherapy*, *6*(6), 722–728. Retrieved from http://www.ncbi.nlm.nih.gov/pmc/articles/PMC444726/.

Hayhurst, E. J., Kailas, L., Hobbs, J. K., & Foster, S. J. (2008). Cell wall peptidoglycan architecture in Bacillus subtilis. *Proceedings of the National Academy of Sciences of the United States of America*, *105*(38), 14603–14608. http://dx.doi.org/10.1073/pnas.0804138105.

Hell, S. W., & Wichmann, J. (1994). Breaking the diffraction resolution limit by stimulated emission: Stimulated-emission-depletion fluorescence microscopy. *Optics Letters*, *19*(11), 780–782. http://dx.doi.org/10.1364/OL.19.000780.

Ikai, A. (2010). A review on: Atomic force microscopy applied to nano-mechanics of the cell. *Advances in Biochemical Engineering and Biotechnology*, *119*, 47–61. http://dx.doi.org/10.1007/10_2008_41.

Johnson, J. W., Fisher, J. F., & Mobashery, S. (2013). Bacterial cell-wall recycling. *Annals of the New York Academy of Sciences*, *1277*, 54–75. http://dx.doi.org/10.1111/j.1749-6632.2012.06813.x.

Kapuscinski, J. (1995). DAPI: A DNA-specific fluorescent probe. *Biotechnic & Histochemistry*, *70*(5), 220–233.

Kawai, Y., Marles-Wright, J., Cleverley, R. M., Emmins, R., Ishikawa, S., Kuwano, M., ... Errington, J. (2011). A widespread family of bacterial cell wall assembly proteins. *The EMBO Journal*, *30*(24), 4931–4941. http://dx.doi.org/10.1038/emboj.2011.358.

Koch, A. (2000). The bacterium's way for safe enlargement and division. *Applied and Environmental Microbiology*, *66*(9), 3657–3663.

Kock, H., Gerth, U., & Hecker, M. (2004). MurAA, catalysing the first committed step in peptidoglycan biosynthesis, is a target of Clp-dependent proteolysis in Bacillus subtilis. *Molecular Microbiology*, *51*(4), 1087–1102.

König, H., Claus, H., & Varma, A. (2010). *Prokaryotic cell wall compounds: Structure and biochemistry*. Springer-Verlag Berlin Heidelberg.

Kühlbrandt, W. (2014). Cryo-EM enters a new era. *eLife*,*3*. e03678.http://dx.doi.org/10.7554/eLife.03678.

Kuru, E., Hughes, V. H., Brown, P. J., Hall, E., Tekkam, S., Cava, F., ... VanNieuwenhze, M. S. (2012). In situ probing of newly synthesized peptidoglycan in live bacteria with fluorescent D-amino acids. *Angewandte Chemie, International Edition*, *51*(50), 12519–12523. http://dx.doi.org/10.1002/anie.201206749.

Kuru, E., Tekkam, S., Hall, E., Brun, Y. V., & VanNieuwenhze, M. S. (2015). Synthesis of fluorescent D-amino acids and their use for probing peptidoglycan synthesis and bacterial growth in situ. *Nature Protocols*, *10*(1), 33–52. http://dx.doi.org/10.1038/nprot.2014.197.

Lairson, L. L., Henrissat, B., Davies, G. J., & Withers, S. G. (2008). Glycosyltransferases: Structures, functions, and mechanisms. *Annual Review of Biochemistry*, *77*, 521–555. http://dx.doi.org/10.1146/annurev.biochem.76.061005.092322.

Lam, H., Oh, D.-C., Cava, F., Takacs, C. N., Clardy, J., de Pedro, M. A., & Waldor, M. K. (2009). D-Amino acids govern stationary phase cell wall remodeling in bacteria. *Science*, *325*(5947), 1552–1555. http://dx.doi.org/10.1126/science.1178123.

Lebar, M. D., Lupoli, T. J., Tsukamoto, H., May, J. M., Walker, S., & Kahne, D. (2013). Forming cross-linked peptidoglycan from synthetic gram-negative lipid II. *Journal of the American Chemical Society*, *135*(12), 4632–4635. http://dx.doi.org/10.1021/ja312510m.

Lecoq, L., Bougault, C., Hugonnet, J. E., Veckerle, C., Pessey, O., Arthur, M., & Simorre, J. P. (2012). Dynamics induced by beta-lactam antibiotics in the active site of Bacillus subtilis L,D-transpeptidase. *Structure*, *20*(5), 850–861. http://dx.doi.org/10.1016/j.str.2012.03.015.

Liechti, G. W., Kuru, E., Hall, E., Kalinda, A., Brun, Y. V., VanNieuwenhze, M., & Maurelli, A. T. (2014). A new metabolic cell-wall labelling method reveals peptidoglycan in Chlamydia trachomatis. *Nature*, *506*(7489), 507–510. http://dx.doi.org/10.1038/nature12892.

Lucic, V., Rigort, A., & Baumeister, W. (2013). Cryo-electron tomography: The challenge of doing structural biology in situ. *Journal of Cell Biology*, *202*(3), 407–419. http://dx.doi.org/10.1083/jcb.201304193.

Lupoli, T. J., Tsukamoto, H., Doud, E. H., Wang, T.-S., Walker, S., & Kahne, D. (2011). Transpeptidase-mediated incorporation of D-amino acids into bacterial peptidoglycan. *Journal of the American Chemical Society*, *133*(28), 10748–10751. http://dx.doi.org/10.1021/ja2040656.

Magnet, S., Bellais, S., Dubost, L., Fourgeaud, M., Mainardi, J.-L., Petit-Frère, S., … Gutmann, L. (2007). Identification of the L,D-transpeptidases responsible for attachment of the Braun lipoprotein to Escherichia coli peptidoglycan. *Journal of Bacteriology*, *189*(10), 3927–3931. http://dx.doi.org/10.1128/JB.00084-07.

Magnet, S., Dubost, L., Marie, A., Arthur, M., & Gutmann, L. (2008). Identification of the L,D-transpeptidases for peptidoglycan cross-linking in Escherichia coli. *Journal of Bacteriology*, *190*(13), 4782–4785. http://dx.doi.org/10.1128/JB.00025-08.

Mahapatra, S., Crick, D. C., & Brennan, P. J. (2000). Comparison of the UDP-N-acetylmuramate: L-alanine ligase enzymes from Mycobacterium tuberculosis and Mycobacterium leprae. *Journal of Bacteriology*, *182*(23), 6827–6830.

Mainardi, J.-L., Fourgeaud, M., Hugonnet, J.-E., Dubost, L., Brouard, J.-P., Ouazzani, J., … Arthur, M. (2005). A novel peptidoglycan cross-linking enzyme for a β-lactam-resistant transpeptidation pathway. *Journal of Biological Chemistry*, *280*(46), 38146–38152. http://dx.doi.org/10.1074/jbc.M507384200.

Matias, V. R. F., Al-Amoudi, A., Dubochet, J., & Beveridge, T. J. (2003). Cryo-transmission electron microscopy of frozen-hydrated sections of Escherichia coli and Pseudomonas aeruginosa. *Journal of Bacteriology*, *185*(20), 6112–6118. http://dx.doi.org/10.1128/jb.185.20.6112-6118.2003.

Matias, V. R., & Beveridge, T. J. (2005). Cryo-electron microscopy reveals native polymeric cell wall structure in Bacillus subtilis 168 and the existence of a periplasmic space. *Molecular Microbiology*, *56*(1), 240–251. http://dx.doi.org/10.1111/j.1365-2958.2005.04535.x.

McCoy, A. J., & Maurelli, A. T. (2006). Building the invisible wall: Updating the chlamydial peptidoglycan anomaly. *Trends in Microbiology*, *14*(2), 70–77. http://dx.doi.org/10.1016/j.tim.2005.12.004.

McIntosh, R., Nicastro, D., & Mastronarde, D. (2005). New views of cells in 3D: An introduction to electron tomography. *Trends in Cell Biology*, *15*(1), 43–51. http://dx.doi.org/10.1016/j.tcb.2004.11.009.

Mengin-Lecreulx, D., Texier, L., Rousseau, M., & van Heijenoort, J. (1991). The murG gene of Escherichia coli codes for the UDP-N-acetylglucosamine: N-acetylmuramyl-(pentapeptide) pyrophosphoryl-undecaprenol N-acetylglucosamine transferase involved in the membrane steps of peptidoglycan synthesis. *Journal of Bacteriology*, *173*(15), 4625–4636.

Mertz, J. (2010). *Introduction to optical microscopy*. Greenwood Village, CO: Roberts and Company Publishers.

Mol, C. D., Brooun, A., Dougan, D. R., Hilgers, M. T., Tari, L. W., Wijnands, R. A., ... Swanson, R. V. (2003). Crystal structures of active fully assembled substrate- and product-bound complexes of UDP-N-acetylmuramic acid: L-Alanine ligase (MurC) from Haemophilus influenzae. *Journal of Bacteriology, 14*, 4152–4162.

Moulder, J. W. (1993). Why is Chlamydia sensitive to penicillin in the absence of peptidoglycan? *Infectious Agents and Disease, 2*(2), 87–99.

Moulder, J. W., Novosel, D. L., & Officer, J. E. (1963). Inhibition of the growth of agents of the Psittacosis group by D-cycloserine and its specific reversal by D-alanine. *Journal of Bacteriology, 85*, 707–711.

Mudd, S., Polevitzky, K., Anderson, T. F., & Chambers, L. A. (1941). Bacterial morphology as shown by the electron microscope: II. The bacterial cell-wall in the genus Bacillus. *Journal of Bacteriology, 42*(2), 251–264.

Mudd, Stuart, Polevitzky, Katherine, Anderson, Thomas F., & Kast, C. C. (1942). Bacterial morphology as shown by the electron microscope: III. Cell-wall and protoplasm in a strain of Fusobacterium. *Journal of Bacteriology, 44*(3), 361.

Murphy, D. B., & Davidson, M. W. (2012). *Fundamentals of light microscopy and electronic imaging* (2nd ed.). New York, NY: John Wiley and Sons, Inc.

Murphy, G. E., Leadbetter, J. R., & Jensen, G. J. (2006). In situ structure of the complete Treponema primitia flagellar motor. *Nature, 442*(7106), 1062–1064. http://www.nature.com/nature/journal/v442/n7106/suppinfo/nature05015_S1.html.

Oliva, M., Dideberg, O., & Field, M. J. (2003). Understanding the acylation mechanisms of active-site serine penicillin-recognizing proteins: A molecular dynamics simulation study. *Proteins, 53*(1), 88–100. http://dx.doi.org/10.1002/prot.10450.

Pilhofer, M., Aistleitner, K., Biboy, J., Gray, J., Kuru, E., Hall, E., ... Jensen, G. J. (2013). Discovery of chlamydial peptidoglycan reveals bacteria with murein sacculi but without FtsZ. *Nature Communications, 4*, 2856.

Pisabarro, A. G., de Pedro, M. A., & Vazquez, D. (1985). Structural modifications in the peptidoglycan of Escherichia coli associated with changes in the state of growth of the culture. *Journal of Bacteriology, 161*(1), 238–242.

Qiao, Y., Lebar, M. D., Schirner, K., Schaefer, K., Tsukamoto, H., Kahne, D., & Walker, S. (2014). Detection of lipid-linked peptidoglycan precursors by exploiting an unexpected transpeptidase reaction. *Journal of the American Chemical Society, 136*(42), 14678–14681. http://dx.doi.org/10.1021/ja508147s.

Real, G., & Henriques, A. O. (2006). Localization of the Bacillus subtilis murB gene within the dcw cluster is important for growth and sporulation. *Journal of Bacteriology, 188*(5), 1721–1732. http://dx.doi.org/10.1128/jb.188.5.1721-1732.2006.

Rust, M. J., Bates, M., & Zhuang, X. (2006). Sub-diffraction-limit imaging by stochastic optical reconstruction microscopy (STORM). *Nature Methods, 3*(10), 793–795. http://dx.doi.org/10.1038/nmeth929.

Sanders, A. N., & Pavelka (2013). Phenotypic analysis of Eschericia coli mutants lacking L,D-transpeptidases. *Microbiology:159*(Pt_9). http://dx.doi.org/10.1099/mic.0.069211-0.

Sauvage, E., Kerff, F., Terrak, M., Ayala, J. A., & Charlier, P. (2008). The penicillin-binding proteins: Structure and role in peptidoglycan biosynthesis. *FEMS Microbiology Review, 32*(2), 234–258. http://dx.doi.org/10.1111/j.1574-6976.2008.00105.x.

Schleifer, K. H., & Kandler, O. (1972). Peptidoglycan types of bacterial cell wall and their taxonomic implications. *Bacteriological Reviews, 36*(4), 407–477.

Sink, R., Barreteau, H., Patin, D., Mengin-Lecreulx, D., Gobec, S., & Blanot, D. (2013). MurD enzymes: Some recent developments. *Biomolecular Concepts, 4*(6), 539–556. http://dx.doi.org/10.1515/bmc-2013-0024.

Sizemore, R. K., Caldwell, J. J., & Kendrick, A. S. (1990). Alternate gram staining technique using a fluorescent lectin. *Applied and Environmental Microbiology, 56*(7), 2245–2247.

Soufo, H. J., & Graumann, P. L. (2003). Actin-like proteins MreB and Mbl from Bacillus subtilis are required for bipolar positioning of replication origins. *Current Biology, 13*(21), 1916–1920.

Stephens, R. S., Kalman, S., Lammel, C., Fan, J., Marathe, R., Aravind, L., … Davis, R. W. (1998). Genome sequence of an obligate intracellular pathogen of humans: Chlamydia trachomatis. *Science, 282*(5389), 754–759.

Stuart Mudd, D. B. L. (1941). Bacterial morphology as shown by the electron microscope: I. Structural differentiation within the streptococcal cell. *Journal of Bacteriology, 41*(3), 415–420.

Tamura, A., & Manire, G. P. (1968). Effect of penicillin on the multiplication of meningopneumonitis organisms (Chlamydia psittaci). *Journal of Bacteriology, 96*(4), 875–880.

Terai, T., & Nagano, T. (2013). Small-molecule fluorophores and fluorescent probes for bioimaging. *Pflügers Archiv : European Journal of Physiology, 465*(3), 347–359. http://dx.doi.org/10.1007/s00424-013-1234-z.

Tiyanont, K., Doan, T., Lazarus, M. B., Fang, X., Rudner, D. Z., & Walker, S. (2006). Imaging peptidoglycan biosynthesis in Bacillus subtilis with fluorescent antibiotics. *Proceedings of the National Academy of Sciences of the United States of America, 103*(29), 11033–11038. http://dx.doi.org/10.1073/pnas.0600829103.

Typas, A., Banzhaf, M., Gross, C. A., & Vollmer, W. (2012). From the regulation of peptidoglycan synthesis to bacterial growth and morphology. *Nature Reviews. Microbiology, 10*(2), 123–136. http://dx.doi.org/10.1038/nrmicro2677.

Ursell, T. S., Nguyen, J., Monds, R. D., Colavin, A., Billings, G., Ouzounov, N., … Huang, K. C. (2014). Rod-like bacterial shape is maintained by feedback between cell curvature and cytoskeletal localization. *Proceedings of the National Academy of Sciences of the United States of America, 111*(11), E1025–E1034. http://dx.doi.org/10.1073/pnas.1317174111.

van Heijenoort, J. (2007). Lipid intermediates in the biosynthesis of bacterial peptidoglycan. *Microbiology and Molecular Biology Reviews, 71*(4), 620–635. http://dx.doi.org/10.1128/mmbr.00016-07.

van Heijenoort, J. (2011). Peptidoglycan hydrolases of Escherichia coli. *Microbiology and Molecular Biology Reviews, 75*(4), 636–663. http://dx.doi.org/10.1128/MMBR. 00022-11.

van Teeseling, M. C., Mesman, R. J., Kuru, E., Espaillat, A., Cava, F., Brun, Y. V., … van Niftrik, L. (2015). Anammox planctomycetes have a peptidoglycan cell wall. *Nature Communications, 6*, 6878. http://dx.doi.org/10.1038/ncomms7878.

Vernon-Parry, K. D. (2000). Scanning electron microscopy: An introduction. *III-Vs Review, 13*(4), 40–44.http://dx.doi.org/10.1016/S0961-1290(00)80006-X.

Verwer, R. W., Nanninga, N., Keck, W., & Schwarz, U. (1978). Arrangement of glycan chains in the sacculus of Escherichia coli. *Journal of Bacteriology, 136*(2), 723–729.

Vollmer, W., Blanot, D., & de Pedro, M. A. (2008). Peptidoglycan structure and architecture. *FEMS Microbiology Review, 32*(2), 149–167. http://dx.doi.org/10.1111/j.1574-6976.2007.00094.x.

Weidenmaier, C., & Peschel, A. (2008). Teichoic acids and related cell-wall glycopolymers in Gram-positive physiology and host interactions. *Nature Reviews. Microbiology, 6*(4), 276–287. http://dx.doi.org/10.1038/nrmicro1861.

Williams, D. B., & Carter, C. B. (Eds.), (1996). The transmission electron microscope. In *Transmission electron microscopy: A textbook for materials science,* Boston, MA: Springer.

Yagi, K. (1987). Reflection electron microscopy. *Journal of Applied Crystallography, 20*(3), 147–160. http://dx.doi.org/10.1107/S0021889887086916.

Yao, X., Jericho, M., Pink, D., & Beveridge, T. (1999). Thickness and elasticity of gram-negative murein sacculi measured by atomic force microscopy. *Journal of Bacteriology, 181*(22), 6865–6875.

Yeats, C., Finn, R. D., & Bateman, A. (2002). The PASTA domain: A β-lactam-binding domain. *Trends in Biochemical Sciences, 27*(9), 438–440. http://dx.doi.org/10.1016/s0968-0004(02)02164-3.

Zapun, A., Contreras-Martel, C., & Vernet, T. (2008). Penicillin-binding proteins and β-lactam resistance. *FEMS Microbiology Review, 32*(2), 361–385. http://dx.doi.org/10.1111/j.1574-6976.2007.00095.x.

Zapun, A., Vernet, T., & Pinho, M. (2008). The different shapes of cocci. *FEMS Microbiology Reviews, 32*, 345–360.

Zernike, F. (1942). Phase contrast, a new method for the microscopic observation of transparent objects. *Physica, 9*(7), 686–698. http://dx.doi.org/10.1016/S0031-8914(42).

CHAPTER 2

Time-lapse microscopy and image analysis of *Escherichia coli* cells in mother machines

Y. Yang*,†,‡, X. Song*,†, A.B. Lindner*,†,‡,1

*INSERM, U1001, Paris, France
†Faculté de Médecine, Université Paris Descartes, Paris, France
‡Center for Research and Interdisciplinarity, Paris, France
[1]Corresponding author: e-mail address: ariel.lindner@inserm.fr

1 INTRODUCTION

Quantitative longitudinal measurements of single cells by time-lapse microscopy have been a popular technique in systems biology and microbial physiology in recent decades. Studies employing these types of experiments have advanced our understanding of cell size control (Campos et al., 2014; Taheri-Araghi et al., 2015), gene expression and regulation (Cai, Dalal, & Elowitz, 2008; Golding, Paulsson, Zawilski, & Cox, 2005; Locke, Young, Fontes, Hernandez Jimenez, & Elowitz, 2011), subcellular organisation and organelle dynamics (Babic, Lindner, Vulic, Stewart, & Radman, 2008; Parry et al., 2014), cellular development and cell fate decision making (Suel, Garcia-Ojalvo, Liberman, & Elowitz, 2006).

However, the exponential nature of microbial growth often limits the number of generations cells can be followed through microscopy. For instance, in experiments where expanding single layers of cells, known as microcolonies, are followed, as they are expanding exponentially over time, tension and friction between cells inevitably push them into multiple layers. Furthermore, the postexperimental image analysis process limits the number of cells whose relevant traits can be quantified. In each image frame, cells have to be segmented from each other. The same cells and their progenies have to be tracked through consecutive frames to form lineages. Consequently, the data generated from a single overnight experiment could take weeks to analyse.

These experimental constraints could be overcome by the applications of microfluidic devices specifically designed to spatially isolate and align single cells or lineages of cells. Examples of such devices are the mother machine, designed to track old pole cells and their immediate progenies of coliform bacteria (Wang et al., 2010), and U-shaped traps (Nagarajan et al., 2014; Rowat, Bird, Agresti, Rando, & Weitz, 2009), designed to isolate mother cells of budding yeasts. These devices contain geometrically fitting microstructures to restrict cells to spatially regular patterns. This geometrical restriction greatly facilitates the task of image analysis. In addition, constant microfluidic flows remove most of the newborn cells and focus the analysis on a constant number of cells within an ever-expanding population.

This chapter aims to use the example of tracking *Escherichia coli* cells in a device known as the mother machine to demonstrate the general principles of such spatially structured microbial time-lapse microscopy experiments. We introduce common methods and potential issues in the design, execution and image analysis of such experiments. Since such experiments involve microfabrication, microscopic, microfluidic and image analysis methods, each of which could be covered in their own chapters, we chose to focus this chapter on the interplay between them.

2 EXPERIMENTAL DESIGNS
2.1 GENERAL PRINCIPLES OF THE MOTHER MACHINE

The mother machine is a good example of microfluidic devices used to isolate and align single bacterial cells in microscopic studies. It was originally designed to track the old pole cells of elongating and dividing lineages of *E. coli*, hence the name. Due to its simplicity of design, ease of use and the ability to track single cells for up 150 generations, its use spreads quickly and has been successfully used to culture and study coliform bacteria such as *E. coli* and *Bacillus subtilis* (Norman, Lord, Paulsson, & Losick, 2013; Wang et al., 2010).

The general principle of the mother machine is very simple: an array of single-cell-wide channels with dead ends on one side, and a much larger main flow channel on the other supplying fresh medium and removing newly born cells. After the cells are loaded into the dead-end channels and fed with fresh medium, they grow into one-dimensional 'colonies'. Once the dead-end channels are completely filled, cells on the open side of the colony will be pushed into the main flow channel and washed away, as shown in Fig. 1A. By constantly supplying fresh medium through the main channel, the growth conditions reach steady states quickly and can be kept in the exponential regime for hundreds of generations (Wang et al., 2010). In these conditions, the dead-end channels contain the inner most cells of the whole lineage, i.e., the mother cells and their immediate progenies.

The obvious benefit of using the mother machine is that cells could be followed for days in exponential growth, without either change in media conditions or obscuration due to overcrowding. This type of long-term longitudinal data could be very useful if the phenomena of interests are rare or take more than 10 generations to develop, such as ageing (Wang et al., 2010) and stochastic developmental decisions (Norman et al., 2013). Another feature of the mother machine and similar microfluidic systems is the ability to easily and quickly shift culturing conditions. Coupled with time-lapse microscopy, single-cell level dynamic behaviour in physiology and gene expression can be easily observed (Izard et al., 2015).

2 Experimental designs

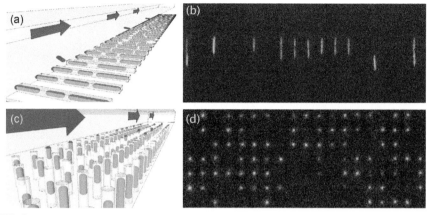

FIG. 1

The general principle and design variations of the mother machine. Shown are the 3D cartoons (A, C) of two variants of the mother machine. Cells are trapped inside dead-end channels and are growing as 1D colonies. Extra cells are pushed outside the dead-end channels and removed by the flow in the main channel. The arrays of dead-end channels are aligned either horizontally (A) or vertically (C) to the imaging plain. (B, D) Actual fluorescent images of these devices containing cells of *E. coli* constitutively expressing a variant of mVenus (shown in *green*). They are also exposed to propidium iodide in the medium, whose fluorescence (shown in *red*) is an indicator of cell death.

One advantage of the mother machine we find especially relevant for this chapter is the fixed locations and geometry of cells and colonies within the mother machine. Instead of having to segment and track hundreds of individual cells from crowded and moving 2D colonies (Ni et al., 2012), one only needs to identify cell divisions and boundaries between cells in one dimension. This made image processing much simpler by reducing manual intervention and error correction, and improving the quantification throughput from time-lapse movies.

The rest of this section discusses in turn the design considerations of both the mother machine itself and time-lapse imaging process. The next section will demonstrate the operational procedure of mother machine time-lapse experiment using our experimental protocol. The last section is dedicated to the image analysis methods of these experiments.

2.2 DEVICE DESIGN CONSIDERATIONS

With a number of reasons for applying the mother machine to time-lapse experiments, the specific objectives of the study determine the exact design and dimensions of the device itself. The first design question to ask is that, in an exponentially expanding lineage of cells, which cells does the experimentalist want to observe? Since the original mother machines will only trap the old pole cells and their immediate progenies, one has to consider whether tracking this subpopulation will either serve the objectives or bias the conclusions. If the objective calls for tracking of different subpopulations, the basic design of the mother machine has to be altered, for example, by opening up the dead ends by connecting them to another channel.

Once the question of which cells within the exponential lineage is settled, the second design decision involves the question of how many cells to follow and the length of the dead-end channels. Dynamic physiological and developmental events often span multiple generations (Izard et al., 2015; Norman et al., 2013), and sometimes correlations between cousins rather than the immediate progeny have to be examined (Hilfinger & Paulsson, 2015). For time-lapse experiments designed to measure these phenomena, the tracked subpopulation has to include every individual within several generations. This requires the dead-end channels to be longer than $L \times 2^N$, where L is the average length of cells and N is the number of generations needed to be followed.

The length of dead-end channels is limited by two factors. The first is the heterogeneity of the growth conditions along the length of dead-end channels. The small cross-sectional area within the channel may limit the diffusion of the media, and the waste resulting from the metabolic activity may accumulate. A variation on the original mother machine designed to relieve diffusion problems (Norman et al., 2013) involves the enlargement of the upper halves of the dead-end channels. The other limiting factor is the surface friction between cells and the device itself, caused by growth and cell displacement. Since one end of the one-dimensional colony is spatially fixed, the other end has to be displaced at a velocity proportional to the total length of the colony and the average growth rate. This friction may cause the misalignment of colonies in a way that is counterproductive to the image analysis process. Thus the dead-end channels should not be longer than necessary. Depending on the requirements of the experiments, growth homogeneity and the correct colony alignment need to be confirmed experimentally (Wang et al., 2010).

The width of the dead-end channels is also important in aligning the cells in ways that are convenient for image analysis and quantification. Specifically, if the dead-end channels are significantly wider than the cells in the intended culture conditions, the friction mentioned earlier may push neighbouring cells into overlapping configurations. Irrespective of whether the overlap is perpendicular to or along the imaging plane, as is the case in Fig. 2, it would significantly complicate the segmentation step in image analysis. Obviously, the dead-end channels should not be too small such as to prevent the cells from entering them. The best approach for determining the correct width of the dead-end channels would be to first measure accurately the cell width under the intended culture conditions, microfabricate a series of mother machines with a range of dead-end channel widths at or around that the measured cell width, and then determine experimentally those with the most suitable width, as described in Section 3.

Last but not the least, different layouts of the main flow channels and dead-end channels allow for either the multiplexing of independent experiments or increasing of the population size in a single device, depending of course on the requirements of the experiment. For multiplexing, the standard approach of having identical units of main channels lying parallel to each other will suffice. This allows independent

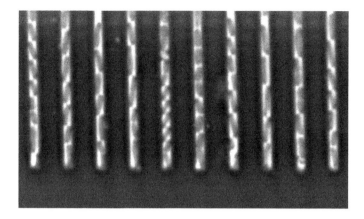

FIG. 2

Phase-contrast image of the dead-end channels of a mother machine filled with narrower *E. coli* cells. Due to the disparities between the widths of the cells and the dead-end channel, the bacterial cells are misaligned and might cause segmentation problems in image analysis.

microfluidic experiments and culture populations be recorded in the same device and time-lapse microscopy session. For the increased population size, Fig. 1C and D demonstrates a variant of the mother machine, where dead-end channels are aligned vertically to the imaging plane and the main flow channel lies on top. This configuration increases the number of cells that can be monitored in a single field of view at the cost of lineage information, and is used in our laboratory to study *E. coli* mother cell lifespan distributions.

2.3 TIME-LAPSE CONSIDERATIONS

In general, time-lapse microscopy can be conducted on mother machines or similar microfluidic devices without major alterations or additional equipments. We conducted all of our mother machine experiments on conventional inverted epifluorescence microscopes from major commercial brands. Besides the microscopes and illumination sources, our platforms are fitted with temperature control chambers, motorised and automatically controlled stages and focusing, fluorescent filter wheels and shutters, and CCDs. The systems are managed by commercial microscope automation software such as MetaMorph® installed on dedicated PCs. Yet, to achieve the best results for time-lapse experiments over long durations, the mother machine is usually employed and, as a result, certain aspects of time-lapse microscopy need to be specially considered during the design process.

An inherent trade-off for multipositional time-lapse microscopy is between the sample size and the time resolution. This trade-off can be expressed as:

$$k_{pos} = \frac{N_{sample}}{\langle n \rangle} \leq \frac{T_{int}}{\langle t_{img} \rangle},$$

where T_{int} is the imaging interval, $\langle t_{img} \rangle$ is the average time of moving to, focusing and imaging one position, $\langle n \rangle$ is the average number of cells or lineages per position, N_{sample} is the total sample size and k_{pos} is the number of imaging positions.

For quantitative experiments aiming to measure dynamic variables such as growth rate and promoter activity, T_{int} should be much shorter than the mean generation time. Even if these measurements are not required, it is necessary to have a minimum of four frames per division time in order to correctly track and segment cells. This poses an upper boundary for T_{int}; and since $\langle n \rangle$ is determined by the feature layout in a microfluidic device and cell loading efficiency, there are also upper boundsaries for k_{pos} and N_{sample}. Before recording the imaging positions and starting the time-lapse loop, the experimentalist should estimate $\langle t_{img} \rangle$ and $\langle n \rangle$ experimentally and set an appropriate k_{pos} and T_{int} regime based on the estimated sample size. If appropriate N_{sample} values cannot be met, aspects of the experiment need to be changed, for example, by increasing the loading efficiency or shortening the focusing and imaging procedure at each position. In order to maximise the overall sample size, it is important to minimise imaging time at each position and only use complex focusing procedures when required, as will be discussed next.

Microfluidic devices, including the mother machine, often complicate the focusing procedures of automatic time-lapse microscopy. This is because the system has to automatically identify the cells within the device among multiple material interfaces and microfabricated structures. Software-based autofocusing methods, such as those maximising image contrasts, may fail because they choose larger and more salient device structures rather than the cells. This problem can be avoided by using hardware-based methods, such as PFS from Nikon or Definite Focus from Zeiss, which use LEDs to track the bottom of the device. However, these systems often require manual resetting when the stage travels large distances and outside their narrow operational z-ranges. In our experience, the most robust and efficient approach is to use software-based autofocusing as a complement to LED-based systems: when they report out-of-range errors, we use contrast maximisation to locate the narrow z-range in which both the microfabricated structures and cells are positioned, thus automatically moving the focus to within the operational range of the hardware systems. This approach can be automated with scripts in MetaMorph® or other microscope automation software packages.

3 EXPERIMENTAL PROCEDURES

In this section, we present our experimental procedures for setting up *E. coli* mother machine experiments for inverted fluorescence microscopy. This protocol is one that is used routinely by nonspecialists in our laboratory. The protocols for these types of experiments are necessarily complex, consisting of several phases, including the preparation of the mother machine, cell cultures, fluidic systems and time-lapse microscopy. Instead of detailing every possible alternative method for each phase of the protocol, we leave the detailed discussions of each phase to specialists. We hope to demonstrate the general process of the experiments with an emphasis on their successful integration.

3.1 MAKING POLYDIMETHYLSILOXANE MOTHER MACHINE DEVICES

We use a soft-lithography approach to fabricate our devices, by casting polydimethylsiloxane (PDMS) structures out of prefabricated master negatives on silicon wafers. The lithography methods used to fabricate these master negatives are outside the scope of this chapter. We only want to emphasise that the choice among different lithography methods depends on the degree of precision required to manufacture the dead-end channels, which in turn depends on natural variations of cell morphologies.

To cast PDMS layers out of the master negatives, we pour uncured PDMS mixes onto the wafers, degas, spread out with either gravity (in case of mm-level design thickness) or spin-coating (in case of μm-level design thickness), and partially cure with heat. We cut out the PDMS slabs with the microfabricated area from this partially cured PDMS and separate it from the wafer. To provide fluidic inlets and outlets for the main fluidic channels, we punch holes in the PDMS slabs with 22-gauge Luer Stubs, entering from the front side where the ends of the main channels are located, and exiting through the backside of slabs. 25-Gauge Luer Stubs are used to remove any PDMS dust particles from the inlets and outlets. Then PDMS layers are bounded with each other or onto cover glasses after both surfaces are treated by room air plasma for 90 s. The bounded device is then immediately fully cured at 80°C overnight. These soft-lithography steps should be done in a clean room or at least in an equivalent compartment in which the air has been cleaned by HEPA filters.

Before mother machines can be used in cell culturing experiments, two types of surface treatment may be necessary. First, due to the small dimensions of the dead-end channels, the natural hydrophobic PDMS surfaces need to be made temporarily hydrophilic to enable the channels to be wetted. Secondly, the main channel surfaces need to be modified to prevent bacterial adhesion and biofilm growth. To achieve both ends, we treat the devices with room air plasma for 90 s and then immediately inject a wetting solution consisting of the cell culture buffer (in our case M9 base medium) and 1.5% polyethylene glycol 400 (PEG 400). Depending on the applications, surfactants such as polysorbate 20 (Tween® 20) can also be added to the wetting solution. As a general rule, before injecting any solutions into the device, including the wetting solution, they need to be prefiltered to prevent blockage by dust particles or crystal sediments. Before it can be loaded with cells, the device needs to be treated with the wetting solution for at least 30 min by maintaining a flow rate of at least 0.2 mL/h.

3.2 SETTING UP THE FLUIDIC SYSTEM

We use pulseless syringe pumps with high precision stepper motors to drive our fluidic systems. Since most of the mother machine applications only require infusions at constant flow rates, pulseless syringe pumps are chosen for their simplicity over alternative solutions such as pressure-based flow control systems. Peristaltic pumps should not be used due to their lack of precision and tendency to generate flow pulses.

To prepare the flow system, we first fill several syringes with the required culture media. If noncovalent surface-modifying molecules such as PEG 400 and Tween® 20 are used in the wetting solution, they should also be added to the culture media to the same concentrations. The media should be filter-sterilised before being loaded into sterile syringes. After the syringes are filled, we also cap them with sterile 0.2-μm syringe filters to prevent cross contamination, before adding 23-gauge blunt-end needles for insertion into tubing. We then load all syringes onto syringe pumps and set the pump to the correct syringe diameters.

We use a semiflexible thin-walled polytetrafluoroethylene (PTFE) tubing (I.D. × O.D. = 0.56 mm × 1.07 mm) to connect the syringes and the microfluidic devices. Traditional rigid PTFE tubing is too inflexible to allow the range of motions needed for multipositional time-lapse microscopy. Yet, the superflexible PVC tubing traditionally used for flexible infusion applications should also be avoided because they have been found to leach phthalate plasticisers into the media (McDonald et al., 2008) that, in our laboratory, was found to have physiological effects on the bacteria. Similar semiflexible plasticiser-free tubing, such as TYGON® formulation 2001, can also be used. 23-Gauge metal couplers are used to connect tubing segments to each other and to the PDMS device.

3.3 CELL CULTURE AND LOADING

Cell cultures can be conducted with conventional methods. Because of their physiological and morphological homogeneity, we usually load harvested exponential phase cells into the mother machine. Successful cell loading requires highly dense cell suspensions, concentrated at least 1000-fold from exponential phase cultures. We usually start with 50-mL culture volumes per main channel. Since the final cultures will be injected into the microfluidic device, we usually filter sterilise culture media to remove dust particles. Once the cultures reach the desired optical densities, cells are harvested by centrifugation at physiological temperatures. If the medium in the mother machine is different from those in exponential cultures, cells should be washed by repeated resuspension and centrifugation.

After the last centrifugation step, minimal volumes of filtered media, usually less than 50 μL, should be used to gently resuspend the cell pellets. Then these dense but homogeneous cell suspensions should be injected into the microfluidic device. The main channels are considered to be completely filled when the cell suspensions are visible in both the inlet and outlet tubing. After the removal of the inlet and outlet tubing and couplers, the device is ready for cell loading by centrifugation. The device should be secured to the centrifuge rotor with the dead-end channels pointing towards the direction of the centrifugal force. In the original mother machine design, cell loading can be done with 15 min centrifugation at $200 \times g$.

Successful cell loading should be confirmed by microscopy. In the ideal case, more than 80% of the dead-end channels should be each filled with at least one cell. If loading is unsuccessful, a mother machine with wider dead ends may be required. If loading is successful, cell suspensions left in the main channel should be washed away immediately to avoid cell adhesion and biofilm formation. In our experience, this is done most efficiently by manually injecting and passing four to six small air packets quickly through each of the main channels. The water-to-air surfaces of the air packets are mostly responsible for removing the cells from the surfaces of the main channels. Passing each air packet should be done in seconds to avoid cells inside the dead-end channels from drying.

3.4 MICROFLUIDIC AND TIME-LAPSE SETUP

The last steps of the protocol involve immobilising the microfluidic device on to the microscope stage while connecting to the fluidic system. We use Scotch tape to fix both the microfluidic device and inlet/outlet tubing onto the microscopy stage adapters, either supplied by the microscope manufactures or custom made. Then we connect the inlet tubing with the syringe pumps and outlet tubing to waste collection. Immediately after, we start media infusion at a high flow rate, usually 2 mL/h but depending on the resistance of the device, to eliminate air packets that form at the tubing connection sites. Once all the media inside the device and the tubing are refreshed, the flow rate is set to those required by the experimental protocol.

Having stabilised the flow system, the time-lapse microscopy automation cycles can be started. We use the Multidimensional Acquisition module of MetaMorph® to manage the automation process. Besides the imaging positions where cells of interests are located, we also add a few positions along the main flow channel to the imaging queue to monitor the overall condition inside the device.

At this stage, the experiments should be left to run on its own. Both temperature and flow rate monitoring can be installed and data logged along with the time-lapse images. The 16-bit images recorded by our CCDs can take up large amount of digital storage space. We have therefore written custom scripts to compress them for uploading onto a centralised data storage facility.

4 IMAGE ANALYSIS FOR LINEAGE CONSTRUCTION AND SINGLE-CELL TRAITS

4.1 PREPROCESSING

The goal of image preprocessing is to enhance the visual appearance and improve manipulations of the data sets. Microfluidic microscopy images often suffer from various difficulties:

- Images may be noisy as a result of limited light intensity.
- Images may suffer from uneven illumination.
- The cell channels may not be aligned vertically.
- The distance between two channels is unknown.

To remove the noise and equalise the illumination, we use a Fourier transformation based band-pass filter on spectral space to eliminate the unwanted high-frequency signals and equalise the background by filtering out the low-frequency signals. Instead of keeping time frames within ImageJ stacks, dead-end channels are cut out of each frame. And then for each individual dead-end channel, frames are displayed

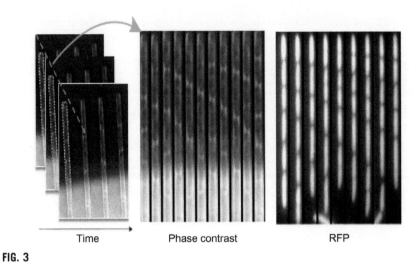

FIG. 3

Time-lapse microscopy images automatically rotated and cut into a time series image.

chronologically from left to right on a single image. This greatly simplifies the tracking process and saves the space needed for restoring the images (see Fig. 3). For example, 500–1000 images can be transformed into 16–19 images that include only cells within dead-end channels. In order to achieve this, the exact distance between two channels in term of pixels and the rotation of the channel should be known. By transforming the edge images obtained by the Canny filter (Canny, 1986) into parameter Hough space (Hough, 1962), the localisations of these lines are defined parametrically within a distribution of points. The average rotation and distance of these lines can be easily computed, based on the most frequently represented points in the transformed parameter space. As a result, rotating, cutting and connecting of the channels can easily be automated.

4.2 SEGMENTATION APPROACHES

It is clear that segmentation accuracy directly affects cell tracking. It therefore makes sense to approach the design of the segmentation method from the point of view of one-dimensional cell sequences inside dead-end channels: The microfluidic device is simpler than the 2D microcolonies, with key information already provided because the cells are restricted to grow in a vertical channel. This type of structure can help to determine what methods of segmentation are truly useful compared to those used to segment 2D microcolonies. With the exemption of simple techniques such as thresholding, segmentation algorithms require the examination of intensity profiles along the centre of the channel.

Based on the type of images that the experiments generate, we will discuss two different segmentation approaches for fluorescent or phase-contrast images. Our results show that the segmentation of fluorescent images can be realised by the dilatation-like regional grow process (Primet, Demarez, Taddei, Lindner, & Moisan, 2008) and the segmentation of phase-contrast image can be implemented on an estimated intensity profile. We introduce these very different approaches and compare results using experimental data.

4.2.1 Phage-contrast image

The benefit of segmenting one-dimensional cell sequences is that segmentation can be realised by peak detection on the intensity profile. In this way, segmentation can act to prevent undesirable results. Instead of detecting the peak on the original image's intensity profile, it is more useful to apply this peak detector to a more textured image, defined by the structure tensor (Bigün & Granlund, 1987). Its purpose is to replace each pixel of the image with their eigenvalues within a predefined scale. Thus, the intensity profile will be extracted from the so-called Eigen image. Fig. 4 clearly shows that the resulting structures are more remarkable and smooth than the original image. The original image's intensity profile caused it to keep a noisey peak and jettison a real peak, simply because the nonuniformed intensity inside the cell. The robustness of this method in decreasing false-positive peak positions and the effect of noise by Eigen image have been confirmed experimentally (see Section 4.4).

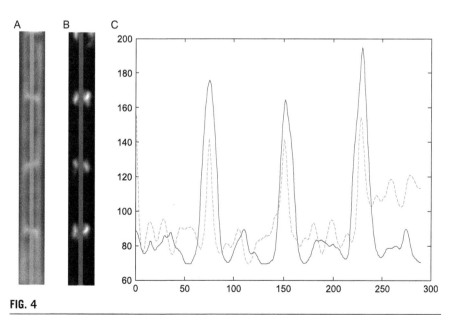

FIG. 4

(A) Original image and (B) structure tensor transformed image. (C) Representation of the intensity profiles through the original image (blue) and Eigen image (green).

4.2.2 Fluorescent image

The use of fluorescent microscopy provides us additional tools to record the time-lapse microscopy images. To capture images of cells in microfluidic devices, fluorescent proteins are often expressed so that cells can be detected with much less noise. As a result, it is important to find a segmentation method based on such images. One of the most powerful tools for doing this is to nonuniformly dilate from predefined seed points in the image and then to recursively add the most similar point as determined by its intensity and geometric distance to the seeds to form labelled regions (Primet et al., 2008). Our challenge is to discover ways to precisely detect the seed points and support the region growing to each seed points, including using automatic local thresholding methods. Although the simple thresholding method is widely used in defining seeds, it lacks precision and consequently the results are limited and no account is taken of geometric connectivity information present of the image. Therefore, we need to carefully reconstruct the seed points' detector from the observed cell intensity distribution. In the method developed by Aguet, Antonescu, Mettlen, Schmid, and Danuser (2013), the point-spread function model is used to configure the statistical p-value of the estimated intensity and the background noise and, as a result, the significance of each candidate signal, detetion sensitivity and selectivity is reinforced over existing single-molecule detection methods. In the fluorescent image shown in Fig. 5, the cell intensity distribution

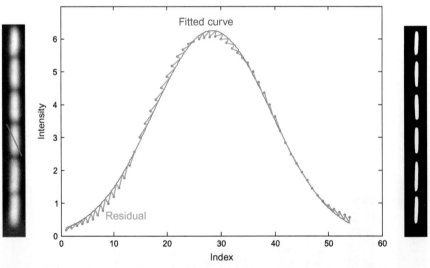

FIG. 5

An example of the intensity profile through a selected cell. The fitted curve is performed by an approximation of Gaussian function. The most statistically significant pixels (seed points) are selected by performing a one-sided, two-sample t-test of the fitted amplitude and the noise, estimated by the residual sum of squares (RSS). The right image presents the mask of the seed points that are defined as $m[k] := p[k] < \alpha$, as described in Aguet et al. (2013).

function is an approximate Gaussian function and the cell intensity profile follows a Gaussian distribution. As a result, the cell intensity can be approximated by the sum of estimated Gaussian amplitude, the background intensity and the noise, as described in Aguet et al. (2013).

4.3 LINEAGE APPROACHES

For time-lapse microscopy images, the most important biological information is the life-histories of individual cells, e.g., cell lineages captured through image analysis. Cell growth normally results in two states of the cell, cell continuation $C \rightarrow$ and cell division $C\%$, each of which needs to be defined as an assignment mapping of a cell at time t and $t+1$ (Fig. 6). Because of the design of the mother machine, a third cell state indicates that a cell is pushed off due to the growth of other cells; cells without identifiable ancestors or descendants are marked as $C \downarrow$. In addition, it is also possible for a cell to explode and disappear; in this case a virtual cell C_v is used to replace the invisible cell to avoid tracking errors. In this section, we will discuss the three main steps for building cell lineage: (1) cell tracking, (2) tracking error detection and (3) semiautomatic error correction.

4.3.1 Cell tracking

To determine the life history of a cell from the first generation until its ultimate fate, cell tracking is required to quantify cell behaviour. Tracking, considered as an assignment model (Jiang, Fels, & Little, 2007), has been studied widely and transformed as a problem of global optimisation. Instead of giving a definite segmentation of a cell, in Jug et al. (2014), the authors supposed to consider all the possible

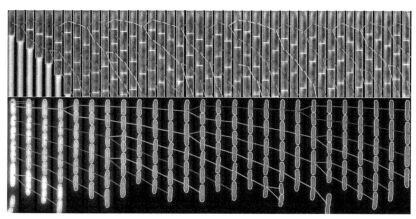

FIG. 6

Cell lineage expressed by linked arrows assigned to cell segmentation calls. The protocol for tracking cells is the same for the phase-contrast and fluorescent images, and therefore no additional segmentation information is needed.

combinations of oversegmentations. Based on the certain physical and geometrical penalties, the most plausable segmentation was chosen via linear programming optimisation, with several constraints related to the simple biology of the system. However, the definition of the penalty, which is at a low level, is incomplete and has developed arbitrarily. Moreover, since the intensity profile must often be determined on the basis of un-delineaged noisy images (see Fig. 4), there is a high risk that the hypothesis of the combined segmentations at the first stage is incomplete. In our work, we are primarily concerned with expressing the tracking observations in a concise assignment hypothesis due to the high accuracy of the segmentation process:

1. $C_{a(t)} \rightarrow C_{b(t+1)}$ (continuation) if the cell $C_{a(t)}$ possessed an ancestor and has not assigned a descendant, meanwhile, if $C_{b(t+1)}$ has not yet an ancestor and if the size of $C_{a(t)} < \lambda_1 C_{b(t+1)}$, then $C_{b(t+1)}$ is assigned as $C_{a(t)}$'s descendant; similarly, $C_{a(t)}$ is assigned as $C_{b(t+1)}$'s ancestor. In our experience, λ_1 is a fixed value that is slightly bigger than 1 ($\lambda_1 = 1.055$) since some cells gradually shrink, especially when starved. In cases where the size of $C_{b(t+1)}$ is beyond the size of $\lambda_2 C_{a(t)}$ ($\lambda_2 = 1.6$), then $C_{b(t+1)}$ will be assigned also to the descendant of the bottom cell (named $C_{a\_bottom(t)}$ if existed) of $C_{a(t)}$ if and only if the size of $C_{a(t)} + C_{a\_bottom(t)} < 1.2 C_{b(t+1)}$. By mapping $C_{b(t+1)}$ to a descendant twice, this assignment reflects an error in segmentation assignment. We use Random Forest, discussed in Section 4.3.2, to detect this kind of error.

2. $C_{a(t)} \% C_{b(t+1)}$ (division) if the cell $C_{a(t)}$ possessed an ancestor and has not assigned a descendant nor a daughter cell; meanwhile, if $C_{b(t+1)}$ has not yet an ancestor and if the size of $C_{a(t)} > \lambda_1 C_{b(t+1)}$, then $C_{b(t+1)}$ is assigned as $C_{a(t)}$'s daughter cell and $C_{a(t)}$ is assigned as $C_{b(t+1)}$'s ancestor. In the case where $C_{b(t+1)}$'s bottom cell existed (named $C_{b\_bottom(t+1)}$), and its size is $1.55 C_{a(t)} < C_{b(t+1)} + C_{b\_bottom(t+1)}$, then $C_{b\_bottom(t+1)}$ is assigned as $C_{a(t)}$'s younger daughter. $C_{a(t)}$ is assigned to $C_{b\_bottom(t+1)}$'s ancestor as well.

3. $C \downarrow$, if from step 1 to step 2 the cell is not assigned as any ancestor or descendant. C_v is assigned to the descendant of the cell if the actual cell has no overlap with the previous cell.

These are effective assignment definitions, not only because the parameters are easy to set up but more importantly because they provide more visual and easy-to-comprehend feedback information. Because these assignment definitions remain largely independent of the cell's image content (either phase contrast or fluorescence) it connects, the segmentation and tracking error detection processes can be performed without requiring the presence of the original images. We will discuss the error detection and correction in the next sections.

4.3.2 Error detection
The approach by which cells are connected by the arrows, in some respects, can influence the identification of segmentation errors: cell tracking with visible linkages such as arrows intuitively reveals the segmentation errors more than the unmarked cells. Nevertheless, to help discriminate between the correct and wrong

segmentations based on these linkages, far more efforts need to be taken. In our work, we make use of a Random Forest classification method (Tin Kam, 1995) to train a classifier of segmentations with respect to its high accuracy and efficiency. By learning the relevant segmentation and tracking information from manually curated training images, Random Forest repeatedly generates decision trees, from bagging to Random Forests, to categorise segmentations based on the weighted trees. It then uses the 'majority vote' to predict the categories of tracking errors. The results showed that the effect of feature selection on the Random Forest was markedly inferior to traditional classification methods. Despite this, it is more appropriate to only give the algorithm the necessary essential features, rather than expecting it to learn to ignore the irrelevant ones. In our work, we have chosen the following measurements as the key features for the Random Forest: For segmentation: mean, standard deviation, minimum value, maximum value, median, mode: for tracking: estimated growth rate, presumed growth rate, difference of cell length and cell centre coordinates, Fourier open curve descriptor; for the descriptor, each cell is considered as a training data point composed of a vector of segmentation and tracking features; for the tracking features, the coefficients of Fourier series were used to describe a lineage around a cell that, in our case, corresponds to an open curve. We adapt the type-P Fourier descriptors using the slope information (Uesaka, 1984) to achieve the invariance of translation, rotation, scale and the start point.

The accuracy of this classification on the test data set was assessed in the context of segmentation classification in Section 4.4. We show that, in most cases, using only tracking descriptors provides very good results compared to segmentation features. Once again that results show that the tracking process helps to detect segmentation errors.

4.3.3 Error correction

Our studies have demonstrated a strong connection between segmentation and tracking. Incorrect segmentation always leads to the wrong assignment of cell lineages. Visibly connecting cells with arrows has greatly increased the detection of segmentation error. In our work, one solution to automatically correcting the segmentation errors, the voting mechanism, is discussed below.

Because of the effort we have put into the segmentation step, around 99.5% of the segmentation calls we make are correct. Incorrect segmentation calls often occur when the cells have exploded, are out of focus, have low light intensity or have aggregates inside. Fortunately, we can use the consensus among the correct segmentation calls to help correct the wrong segmentation calls by what we call the 'voting mechanism'. The ancestor and descendant of a wrong segmentation call will be all considered as a voting part. The majority vote taken by this mechanism will decide if the cell is over- or undersegmented and this process is illustrated in Fig. 7. The voting mechanism benefits from the high accuracy of the correct segmentation calls so that their participation in the voting procedure greatly increases the judgement of wrong segmentation calls.

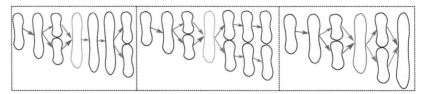

FIG. 7

An illustration of the voting mechanism. There are two states of wrong segmentation calls: oversegmentation and undersegmentation. The voting mechanism is used to determine which type of segmentation error is more likely to be the case considering all the evidence present in the time lapse frames, and then correct the errors accordingly. The voting score is determined by the difference in the number of descendants and the number of ancestors within the same generation. If the measured score is bigger than 0, then the ancestor should be merged with the younger sister. If the score is less than 0, the wrong segmentation call is divided into two segmentations. If the score equals 0, no action is to be taken until intervention from the operator. From *left* to *right*, the difference in the voting score is: $(3-1)>0$, $(1-4-n)<0$ and $(1-1)=0$. Therefore, the ancestor of the red segmentation is merged with its bottom cell. The *green segmentation* is split into two cells. For the *blue segmentation*, no correction is taken. The voting score is computed iteratively because the value will be changed after every correction process.

4.4 IMAGE ANALYSIS PERFORMANCES AND RESULTS

We propose to assess segmentation power and tracking accuracy through two types of images, phase contrast and fluorescent. To this end, we performed five data sets for each type of image to evaluate the performance of segmentation and tracking. Each data set contains 101–280 images, each image contains 17–19 channels and each channel contains 4–6 cells. Hence, one position contains at least $101*17*4$ (8k) cells. The accuracy of the segmentation and tracking is computed on these data sets. We will also discuss the segmentation error detection rate by Random Forest and their correction ability by the voting mechanism. The software is written as a plugin of ImageJ.

Fig. 8 represents the Recall–Precision curve for different imaging conditions across increasing variations of the original image. Each plot corresponds to a Recall–Precision curve of our segmentation algorithm performed on the image transformed from the original image. For evaluating the robustness of our algorithms, we manually change the imaging condition by adding Gaussian noise and Gaussian blur. The segmentation process labels examples of cells as either positive or negative. The result can be represented in a structure as a confusion matrix that has four categories: True positives (TP) are cell segmentations correctly labelled as positive. True negatives (TN) refer to the wrong segmentations correctly labelled as negative. False positives (FP) correspond to the wrong cell segmentations incorrectly labelled as positive. Finally, the false negatives (FN) are the good cell segmentations labelled as negative.

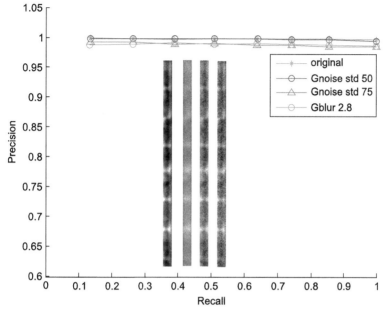

FIG. 8

In the graph, the four Recall–Precision curves represent the performance of our segmentation algorithms. For evaluating robustness, the original image was transformed into images with a *white Gaussian noise* (standard deviation from 50 to 70) and into a blurred image with a Gaussian smoothing by scale 2.8. The used data set contains 101 images with 17 channels. Therefore, around 8k (101*17*5) cells were tested.

	Positive	Negative
True	TP	TN
False	FP	FN

The recall and precision are obtained from the formula $Recall = \frac{TP}{TP+FN}$, $Precision = \frac{TP}{TP+FP}$. In our work, we notice that a very high recall rate was obtained even through very difficult imaging conditions. This means that the average missed cell segmentation rate is around 0.07%. In addition, the resulting precision rate confirms the high accuracy rate of cell segmentation recognition. Compared with conventional methods, the results show that the proposed method has advantages of a high and robust ability to call segmentations correctly, and can be used for the various imaging conditions. For the fluorescence image, the mean accuracy rate of correct cell segmentation calls is around 99.97%, based on the analysis of 5 data sets with 361 images. The background illumination is often nonuniformed because of the sensitivity of fluorescence, so we do not alter the imaging conditions manually.

Despite the fact that the segmentation process gives a highly correct accuracy rate of cell recognition, we still get some erroneous segmentation calls due to noise, uneven illumination and cell aggregation, and therefore error identification is still necessary. This step is performed by Random Forest which is extracted from the application programming interface of a java machine learning package WEKA (Smith & Frank, 2016). The training feature vector is composed of the four tracking descriptors and eight Fourier coefficients, as described in Section 4.3.2. In our work, we use 10% correct segmentations and 30% error segmentations as positive and negative examples to train Random Forest. We obtained a recall rate of around 99.97% and a precision rate of around 99.85%. This means that our detection algorithm is able to determine the category of the extracted features in predicting the segmentation error. Cell segmentation error correction is implemented by the voting mechanism that allows the neighbour segmentation to vote for the forthcoming action. In our work, we successfully corrected 71.5% of the segmentation errors.

4.5 IMAGE ANALYSIS SUMMARY

We have developed an approach for microfluidic bacterial automatic analysis of microscopy image sequences. This approach starts from preprocessing to the cell recognition. We use Hough Transform to detect the channel rotation and the internal distance between two channels. Based on the customised structures of the microfluidic devices, a robust tensor structure-based segmentation method is used for the segmentation of phase-contrast image. We also introduce a statistical analysis, based on the point-spread function to detect the seeds for cell segmentation of fluorescence images. Lineage information is then put into the learning system. Although a high precision rate of segmentation was obtained from the proposed method, there is still a need to detect the erroneous segmentation calls. Using the features of Fourier open curve coefficients, we successfully detected cell segmentation errors that were subsequently corrected using a new voting mechanism algorithm.

ACKNOWLEDGMENT

This work was supported by the AXA Foundation Longevity Chair and the French National Research Agency (ANR) grant.

REFERENCES

Aguet, F., Antonescu, C. N., Mettlen, M., Schmid, S. L., & Danuser, G. (2013). Advances in analysis of low signal-to-noise images link dynamin and AP2 to the functions of an endocytic checkpoint. *Developmental Cell*, *26*(3), 279–291. http://dx.doi.org/10.1016/j.devcel.2013.06.019.

Babic, A., Lindner, A. B., Vulic, M., Stewart, E. J., & Radman, M. (2008). Direct visualization of horizontal gene transfer. *Science*, *319*(5869), 1533–1536. http://dx.doi.org/10.1126/science.1153498.

Bigün, J., & Granlund, G. H. (1987). Optimal orientation detection of linear symmetry. In *Proceedings of the IEEE first international conference on computer vision, 6* (pp. 433–438).

Cai, L., Dalal, C. K., & Elowitz, M. B. (2008). Frequency-modulated nuclear localization bursts coordinate gene regulation. *Nature, 455*(7212), 485–490. http://dx.doi.org/10.1038/nature07292.

Campos, M., Surovtsev, I. V., Kato, S., Paintdakhi, A., Beltran, B., Ebmeier, S. E., & Jacobs-Wagner, C. (2014). A constant size extension drives bacterial cell size homeostasis. *Cell, 159*(6), 1433–1446. http://dx.doi.org/10.1016/j.cell.2014.11.022.

Canny, J. (1986). A computational approach to edge detection. *IEEE Transactions on Pattern Analysis and Machine Intelligence, PAMI-8*(6), 679–698. http://dx.doi.org/10.1109/TPAMI.1986.4767851.

Golding, I., Paulsson, J., Zawilski, S. M., & Cox, E. C. (2005). Real-time kinetics of gene activity in individual bacteria. *Cell, 123*(6), 1025–1036. http://dx.doi.org/10.1016/j.cell.2005.09.031.

Hilfinger, A., & Paulsson, J. (2015). Systems biology: Defiant daughters and coordinated cousins. *Nature, 519*(7544), 422–423. http://dx.doi.org/10.1038/nature14210.

Hough P. V. C. (1962). Method and means for recognizing complex patterns. Google Patents.

Izard, J., Gomez Balderas, C. D., Ropers, D., Lacour, S., Song, X., Yang, Y., ... de Jong, H. (2015). A synthetic growth switch based on controlled expression of RNA polymerase. *Molecular Systems Biology, 11*(11), 840. http://dx.doi.org/10.15252/msb.20156382.

Jiang, H., Fels, S., & Little, J. J. (2007). A linear programming approach for multiple object tracking. In *Paper presented at the 2007 IEEE conference on computer vision and pattern recognition, 17–22 June 2007*.

Jug, F., Pietzsch, T., Kainmüller, D., Funke, J., Kaiser, M., van Nimwegen, E., ... Myers, G. (2014). Optimal joint segmentation and tracking of *Escherichia coli* in the mother machine. In M. J. Cardoso, I. Simpson, T. Arbel, D. Precup, & A. Ribbens (Eds.), *Bayesian and graphical models for biomedical imaging: First international workshop, BAMBI 2014, Cambridge, MA, USA, September 18, 2014, revised selected papers* (pp. 25–36). Cham, Switzerland: Springer International Publishing.

Locke, J. C., Young, J. W., Fontes, M., Hernandez Jimenez, M. J., & Elowitz, M. B. (2011). Stochastic pulse regulation in bacterial stress response. *Science, 334*(6054), 366–369. http://dx.doi.org/10.1126/science.1208144.

McDonald, G. R., Hudson, A. L., Dunn, S. M., You, H., Baker, G. B., Whittal, R. M., ... Holt, A. (2008). Bioactive contaminants leach from disposable laboratory plasticware. *Science, 322*(5903), 917. http://dx.doi.org/10.1126/science.1162395.

Nagarajan, S., Kruckeberg, A. L., Schmidt, K. H., Kroll, E., Hamilton, M., McInnerney, K., ... Rosenzweig, F. (2014). Uncoupling reproduction from metabolism extends chronological lifespan in yeast. *Proceedings of the National Academy of Sciences of the United States of America, 111*(15), E1538–1547. http://dx.doi.org/10.1073/pnas.1323918111.

Ni, M., Decrulle, A. L., Fontaine, F., Demarez, A., Taddei, F., & Lindner, A. B. (2012). Pre-disposition and epigenetics govern variation in bacterial survival upon stress. *PLoS Genetics, 8*(12). e1003148. http://dx.doi.org/10.1371/journal.pgen.1003148.

Norman, T. M., Lord, N. D., Paulsson, J., & Losick, R. (2013). Memory and modularity in cell-fate decision making. *Nature, 503*(7477), 481–486. http://dx.doi.org/10.1038/nature12804.

Parry, B. R., Surovtsev, I. V., Cabeen, M. T., O'Hern, C. S., Dufresne, E. R., & Jacobs-Wagner, C. (2014). The bacterial cytoplasm has glass-like properties and is fluidized by metabolic activity. *Cell, 156*(1–2), 183–194. http://dx.doi.org/10.1016/j.cell.2013.11.028.

Primet, M., Demarez, A., Taddei, F., Lindner, A. B., & Moisan, L. (2008). Tracking of cells in a sequence of images using a low-dimension image representation. In *Paper presented at the 2008 5th IEEE international symposium on biomedical imaging: From nano to macro, 14–17 May*.

Rowat, A. C., Bird, J. C., Agresti, J. J., Rando, O. J., & Weitz, D. A. (2009). Tracking lineages of single cells in lines using a microfluidic device. *Proceedings of the National Academy of Sciences of the United States of America*, *106*(43), 18149–18154. http://dx.doi.org/10.1073/pnas.0903163106.

Smith, T. C., & Frank, E. (2016). Introducing machine learning concepts with WEKA. In E. Mathé & S. Davis (Eds.), *Statistical genomics: Methods and protocols* (pp. 353–378). New York, NY: Springer.

Suel, G. M., Garcia-Ojalvo, J., Liberman, L. M., & Elowitz, M. B. (2006). An excitable gene regulatory circuit induces transient cellular differentiation. *Nature*, *440*(7083), 545–550. http://dx.doi.org/10.1038/nature04588.

Taheri-Araghi, S., Bradde, S., Sauls, J. T., Hill, N. S., Levin, P. A., Paulsson, J., ... Jun, S. (2015). Cell-size control and homeostasis in bacteria. *Current Biology*, *25*(3), 385–391. http://dx.doi.org/10.1016/j.cub.2014.12.009.

Tin Kam, H. (1995). Random decision forests. In *Proceedings of the third international conference on document analysis and recognition, 14–16 August*.

Uesaka, Y. (1984). A new Fourier descriptor applicable to open curves. *Electronics and Communications in Japan (Part I: Communications)*, *67*(8), 1–10. http://dx.doi.org/10.1002/ecja.4400670802.

Wang, P., Robert, L., Pelletier, J., Dang, W. L., Taddei, F., Wright, A., & Jun, S. (2010). Robust growth of *Escherichia coli*. *Current Biology*, *20*(12), 1099–1103. http://dx.doi.org/10.1016/j.cub.2010.04.045.

CHAPTER

Microfluidics for bacterial imaging

3

L.E. Eland*,†, A. Wipat*,†, S. Lee†, S. Park*,†, L.J. Wu†,1

*Interdisciplinary Computing and Complex BioSystems Research Group, School of Computing Sciences, Newcastle University, Newcastle upon Tyne, Tyne and Wear, United Kingdom
†Centre for Bacterial Cell Biology, Institute for Cell and Molecular Biosciences, Newcastle University, Newcastle upon Tyne, Tyne and Wear, United Kingdom
1Corresponding author: e-mail address: l.j.wu@newcastle.ac.uk

1 INTRODUCTION

The past two decades have seen great advances in microscopy technology, especially the development of microscopes with high and 'super' (subdiffraction limit) resolutions. Complementing the advent of high-power microscopes is the availability of new or improved methods for labelling cellular targets. In the field of microbial research, there is an ever-increasing need for live cell imaging and long-term examination of living cells in order to observe cellular processes and biological machinery in action, and to track changes at the single-cell level as well as at a multicellular/population scale.

One of the commonly used methods for live bacterial cell imaging is to mount bacteria on agarose pads, which helps to immobilise the cells. Agarose pads infused with culture media will also support bacterial growth for a certain period of time. However, over time nutrients and oxygen will become limited, or the agarose pads dry out. Furthermore, many studies require growth media to be changed, or bacteria to be free swimming. The application of microfluidics to bacterial imaging not only resolves some of these issues but also opens new avenues for research, by creating and maintaining microenvironments to meet the specific needs of experiments or bacteria.

Microfluidics is the science and technology of manipulating fluid flow in geometrically constrained spaces, usually networks of channels and shaped chambers, at a microscale (Whitesides, 2006; Whitesides, Ostuni, Takayama, Jiang, & Ingber, 2001). The field emerged in the 1980s, when microelectromechanical systems (MEMS), tiny integrated devices or systems that combine mechanical and electrical components, were first manufactured. Subsequently, these were used in industries such as in pressure sensor and inkjet printer head manufacturing. In the 1990s MEMS started to be applied to the fields of biology, biomedicine and chemistry, initially for analysis and then also for detection and other applications such as drug delivery, further fuelling the development of microfluidics. The best known are the Lab-on-a-Chip systems, sometimes also called Bio-MEMS or miniaturised total analysis systems (μTASs), which are miniaturised devices that integrate one or several laboratory procedures on a single chip of submillimetre scale (Volpatti & Yetisen, 2014).

The field of microfluidics involves several disciplines including engineering, physics, chemistry, computer science, nanotechnology, biology and medicine. Microfluidics can reduce sample and reagent consumption and increase automation and parallelisation, helping to reduce cost as well as analysis time (Mark, Haeberle, Roth, von Stetten, & Zengerle, 2010). The past decade has seen a great increase and expansion in microfluidic applications, spanning from manufacturing, diagnostics, analysis, to research. In biology, microfluidic technology is widely used in genomic and proteomic analyses, as well as flow cytometry, detection and water analysis (Beebe, Mensing, & Walker, 2002; Bennett & Hasty, 2009; Liu & Singh, 2013; Streets & Huang, 2014; Velve-Casquillas, Le Berre, Piel, & Tran, 2010; Whitesides, 2006; Whitesides et al., 2001).

One field that has benefited immensely from the development of microfluidics is the microscopic study of microorganisms, which is being revolutionised with very positive impacts (Hol & Dekker, 2014; Saleh-Lakha & Trevors, 2010; Weibel, Siegel, Lee, George, & Whitesides, 2007; Wessel, Hmelo, Parsek, & Whiteley, 2013). The technology allows bacteria to be grown in chemically and physically controlled environments, monitored in situ, in a noninvasive manner, while generating automated, real-time data on the behaviour and morphology of the cells with single-cell resolution (Balagaddé, You, Hansen, Arnold, & Quake, 2005).

One of the earliest bacterial studies using microfluidic microscopy was reported in 2004, where Balaban, Merrin, Chait, Kowalik, and Leibler (2004) created a microfluidic device with channels in which cells were forced to grow, thereby constraining the movement of the daughter cells and allowing the lineage of each cell to be traced. Furthermore, fluid flow in the device provided the cells with media and antibiotics in a controlled manner. Using this technique the authors were able to identify persister cells and show that persistence was linked to preexisting heterogeneity in bacterial populations. Examples of further applications of microfluidic microscopy in bacterial research are listed in Table 1.

While microfluidics has presented new opportunities for bacterial imaging, the broadening application of microfluidics in biological research is also driving rapid development of the technology. Several commercial microfluidics systems for microscopy are now available and have relatively standard protocols defined by the manufacturers (e.g. the BioFlux system and the CellASIC System). However, many applications will require specially designed microfluidic setups tailored for specific applications. In this chapter, we outline the general procedures involved in fabricating, setting up and controlling microfluidic devices (Sections 2–5) and review some recent microfluidic applications in bacteria imaging, focusing on research in the engineering of bacterial systems (Section 6), multispecies biofilm research

Table 1 Examples of Microfluidic Microscopy in Bacterial Research

Application	References
Persistence and persisters	Balaban et al. (2004), Wakamoto et al. (2013), Maisonneuve and Gerdes (2014)
Chemotaxis	Ahmed, Shimizu, and Stocker (2010), Rusconi, Guasto, and Stocker (2014), Chen et al. (2015)
Bacteria detection	Gómez-Sjöberg, Morisette, and Bashir (2005), Boedicker, Li, Kline, and Ismagilov (2008)
Single-cell gene expression	Longo and Hasty (2006), Taniguchi et al. (2010)
Gene expression regulation	Wallden and Elf (2011)
Chromosome behaviour	Larson et al. (2006), Pelletier et al. (2012)
Bacterial interactions	Kim, Boedicker, Choi, and Ismagilov (2008), Wessel et al. (2013)
Motility	Kaehr and Shear (2009), Ducret, Théodoly, and Mignot (2013)
Control of bacterial development	Norman, Lord, Paulsson, and Losick (2013)
Bacterial physiology	Campos et al. (2014), Taheri-Araghi, Brown, Sauls, McIntosh, and Jun (2015)
Biofilm formation	Nance et al. (2013), Kolderman et al. (2015), Oliveira et al. (2015)
Quorum sensing	Boedicker, Vincent, and Ismagilov (2009), Jeong, Jin, Lee, Kim, and Lee (2015)
Stress adaptation	Wang et al. (2014), Mathis and Ackermann (2016)
Identifying new bacterial species	Williams et al. (2016)
Single-molecule based super-resolution microscopy (smSRM)	Cattoni, Fiche, Valeri, Mignot, and Nöllmann (2013)
Host–pathogen interactions	Kim, Hegde, and Jayaraman (2010), Pamp, Harrington, Quake, Relman, and Blainey (2012)

and microbial ecology (Section 7), bacterial cell cycle and size homeostasis (Section 8) and cell shape and geometry (Section 9). Many of these applications used custom-designed microfluidic devices. We also provide step-by-step methods for two of these noncommercial devices (Section 11), both relatively easy to follow and adopt, with the aims of supplying practical information as well as to provoke ideas that can lead to new or further improved microfluidic devices.

2 FABRICATION OF MICROFLUIDIC DEVICES

Though there are a range of commercially available microfluidics devices available, there are many advantages to setting up a bespoke system. Although it takes time, the benefits of being able to design, build, test, feedback and improve the microfluidics devices are significant. This 'do-it-yourself' approach means that such devices can be more easily tailored to address specific research questions. Of course, there are important considerations that are required to integrate microfluidics with microscope systems, such as the working distance and focal distance from the objective, through the base of the device and into the channels containing the cells.

A microfluidic chip is part of a microfluidic device in which microscaled features (microchannels and chambers) have been moulded, and therefore the most important component of the device. To avoid confusion, in this chapter we refer to the assembly that we mount on the microscope stage, mainly consisting of the microfluidic chip, the glass slide(s)/dish and any other accessories that form and support the channels and the chambers, as a microfluidic 'device'. And the 'device' connected with the fluid controller and the microscope as a microfluidic 'setup'. The basic steps of a typical process for fabricating a microfluidics chip, using soft lithography to produce a mould and then replica moulding with polydimethylsiloxane (PDMS) for the chip, can be seen in Fig. 1. Replica-moulded PDMS microfluidics chips are the predominant type of device used for bacterial applications (McDonald et al., 2000). Some of the elements of this process require specialised equipment. However, it is possible to simplify the process by outsourcing some of the more complex process steps, such as lithography of the silicon wafer and the production of the mask.

A number of compact and contained systems for the design and cost-effective production of microfluidics devices for rapid prototyping are also being developed (e.g. Boston University's CIDAR lab). Methods vary but include the 3D printing of polymers (Dolomite Ltd., Royston, UK) or direct milling of hard plastic substrates using a CNC (computer numerical control milling machine) (e.g. MakerFluidics from CIDAR Lab, Boston University). However, most of these systems are limited to a minimum feature size of 5 μm or more depending on the system employed.

2.1 DESIGN

A number of design software options are available but typically Tanner L-edit software (Mentor Graphics, OR, USA) or AutoCAD software (AutoDesk, USA) is used to produce two-dimensional representations of the channel and chamber layout. The design will first need to be transferred onto a mask, for use later in the soft lithography stage. Masks are available from a number of companies, but require specific file formats, typically GDSII or CIF, and this needs to be checked before choosing the design software.

There are a number of useful guides available on the websites of some of the leading microfluidics laboratories on how to design the features in your microfluidics chip. The Stanford microfluidics foundry website (http://web.stanford.edu/group/foundry/) provides a wealth of useful design tips, as well as guides for using design software and some helpful design files.

As an alternative to traditional CAD software, a number of laboratories are developing design tools to assist with various parts of the design process, including Microfluidics design kit (PhoeniX software, The Netherlands) and Fluigi (Huang, 2015). The focus of these is to automate, speed up and simplify the process of designing functioning microfluidics.

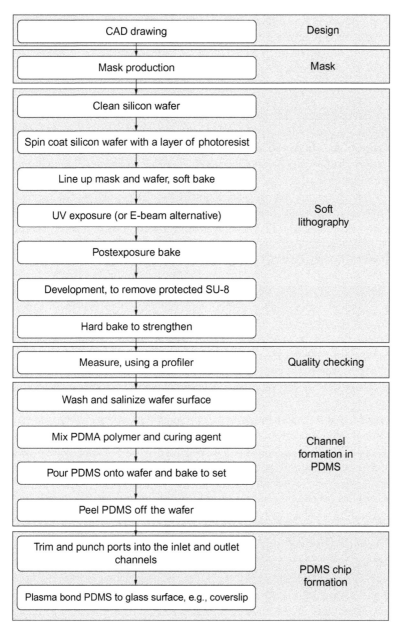

FIG. 1

Flow chart of the main stages involved in producing a PDMS microfluidics chip, using soft lithography and replica moulding.

One of the key considerations when designing microfluidics channels and chambers is the aspect ratio, which is the ratio of height to width. Channels with an aspect ratio of less than 1:10 are likely to fail, with the channel collapsing in onto the bottom surface of the chip during bonding. Too large an aspect ratio (>1:1) can lead to failure in the lithography methods. These general rules vary according to the photoresist or production methods to be used, so it is important to check the specifications of the processes available at the beginning of the design process.

When the microfluidics designs involve two or more depths, regardless of when the two layers need to be aligned, two masks are needed. Both masks should include alignment marks, usually crosses incorporated at the design state, to facilitate their accurate alignment.

In order to design chips with the required fluid flow regime, the movement of liquid through microscale channels needs to be understood. Fluid dynamics at the scales relevant to microfluidics are markedly different to that at larger scales, with laminar flow and minimal mixing dominating. Laminar flow occurs when the Reynolds number is low (typically less than 1000), with the Reynolds number being defined as the inertial forces divided by the viscous forces acting within the channel. More detailed information on fluid flow in microfluidics can be seen in Sharp, Adrian, Santiago, and Molho (2005) and the Elveflow website (http://www.elveflow.com/microfluidic-tutorials/).

At this stage, before going into production, it is useful to model the flow regime and the behaviour of particles, nutrients, chemical concentrations and cells within the design. This ensures that the design is fit for purpose and allows the design to be modified if required preproduction. Modelling can be done with a number of software packages, such as COMSOL Metaphysics and FLOW-3D (FlowScience Inc., USA). COMSOL has a microfluidics package designed specifically to deal with flow through microscale structures. Westerwalbesloh et al. (2015) created a model using COMSOL, which enabled them to assess nutrient limitation in their microfluidics chip design. COMSOL has also been used by the authors of this chapter to model the fluid dynamics in a quorum sensing microfluidics assay chip, and the concentration of an inducer molecule applied to the chip through one of the inlets (Fig. 2).

2.2 PRODUCTION OF SILICON WAFERS

As seen in Fig. 1, photolithography with ultraviolet light is frequently used to activate cross-linking in the photoresist compound beneath. However, the resolution and therefore the minimum feature size that can be achieved are limited by the wavelength of light used. When higher resolution is required for resolving smaller features, more expensive technologies, such as electron beams (E-beam) or ion beams, need to be used. This technology has been utilised for making microfluidics chips for the trapping and study of submicron-sized cells (Moolman, Huang, Krishnan, Kerssemakers, & Dekker, 2013).

An alternative to E-beam lithography is to use a standard photolithographic process and then use surface deposition to deposit a thin and even layer of a compound

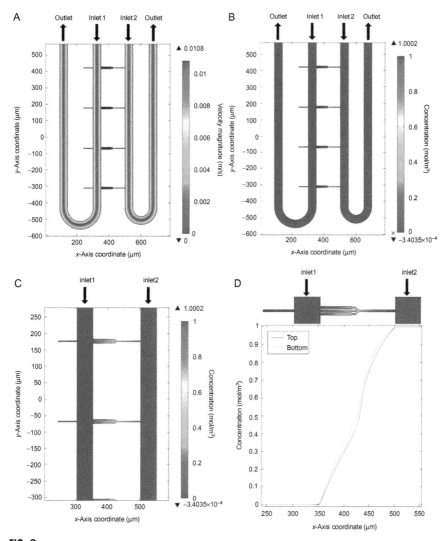

FIG. 2

COMSOL simulation of fluid dynamics in a quorum sensing microfluidics assay chip, designed at Newcastle University by S. Park. (A) The liquid velocity profile of the fluidic channels given 70,000 Pa of pressure on the two inlets. (B) The concentration gradient of the fluidic channels when water is injected through inlet 1 and 1 mM concentration of inducer molecule is injected through inlet 2. (C) A close-up view of the concentration gradient field shown in (B). (D) Comparison of the concentration gradient across the a-axis between two locations (*Top* and *Bottom*) separated by about 6000 μm in the y-axis direction.

such as silicon oxide onto the surface, making the features smaller. This improves resolution so that channels can be made small enough to hold individual bacteria in a constrained position (Moffitt, Lee, & Cluzel, 2012) but without the cost of the E-beam methods.

The complexity of the channels and chambers required will determine the methods chosen for chip and silicon wafer fabrication. For example, if a chip has a number of features and all are the same depth, then a straightforward photolithography method using a layer of photoresist on the silicon surface is all that will be required. For more complex designs, where channels of different depths are required (e.g. where channels of a few microns depth lead off a deeper channel through which media flows), different methods need to be employed. There are two approaches to resolve this fabrication problem. The first option is to use photolithography to create the shallow channels in a photoresist layer and then etch again down into the silicon surface of the wafer itself for the deeper channels. The second option is to make two silicon wafers, using two separate masks: one mask for the shallow channels and the other for the deeper channels. The two layers are then replica moulded in PDMS at the required thickness, and the two layers aligned and bonded to produce the required chip. Alignment can be a challenge, as precision is required. This has led to the development of low cost in-house solutions to align wafers, such as that used by Li et al. (2015).

As mentioned earlier, the design needs to be translated onto the silicon wafer from which the PDMS microfluidic chip will be moulded (Fig. 1). This is done using an intermediate called a mask. Masks are typically made from quartz or soda-lime glass with a chromium surface or from celluloid film. Celluloid film is by far the cheapest mask material, but are less hard wearing and can only be used reliably for design features above 5 µm. On the other hand, glass masks are more expensive but can be used repeatedly with little degradation and for features down to approximately 0.8 µm. Both of these photomask types are suitable for use with UV photolithography.

Another important consideration for the transfer of a design onto a silicon wafer coated in a photoresist resin is whether the polymer used is a negative or positive photoresist, and how the chips will be moulded from it. A negative photoresist is a photoresist in which areas of the resist that are exposed to UV become cross-linked, and in all protected areas the photoresist can be washed away. When using a negative photoresist the mask is designed so that the area that will form the channel is transparent and the rest of the area is opaque (either by the printed or the chrome part of the mask). A commonly used negative photoresist is SU-8 (MicroChem Inc. and Gersteltec SARL), which can be used for the reliable production of features with a depth of between 1 µm and 1 mm. SU-8 is deposited onto the surface of the silicon wafer by centrifugation, with the rotational speed and time determining the thickness of the layer, which in turn determines the final depth of the microfluidic channel that is moulded from the wafer. New formations of SU-8 include SU-8 2000 (MicroChem Inc.; http://www.microchem.com/pdf/SU-82000DataSheet2025thru2075Ver4.pdf), which can be coated to make a minimum layer thickness of 0.5 µm, and SU-8

3000 (MicroChem Inc.; http://microchem.com/pdf/SU-8%203000%20Data%20Sheet.pdf) which can only be coated to 4 µm, but can be used to make structures with aspect ratios up to 5:1.

2.3 PRODUCTION OF THE MICROFLUIDICS CHIP

The next stage in the production of a microfluidics chip is to translate the microscale features from the silicon wafer onto the final chip material. Chips can be made from a range of materials including PDMS, SU-8 (Sikanen et al., 2007), agarose gel (Moffitt et al., 2012), glass and even paper (Tao, Xiao, Lei, & Lee, 2015). However, agarose and PDMS are most commonly used for bacterial applications, in conjunction with the soft lithography methods for patterning wafers. PDMS is an excellent material with which to produce microfluidic devices. It is inexpensive, optically transparent (down to approximately 280 nm) and has low autofluorescence (Piruska et al., 2005). PDMS is also nontoxic to microorganisms and is chemically inert (Mata, Fleischman, & Roy, 2005). PDMS can be used to reproduce submicron-sized features with high fidelity, so is ideal for its use in replica moulding from a patterned silicon-SU-8 wafer. PDMS is relatively easy to work with in most biological laboratories as it cures at low temperatures and is able to be covalently bonded to itself or other silica-based materials after plasma activation (McDonald et al., 2000). Due to its elastomeric properties, PDMS forms smooth surfaces, a feature that is very difficult to achieve when fabrication devices from harder and more brittle materials such as glass or acrylic. The flexibility and elastomeric properties of PDMS also make the formation of holes for inlet and outlet ports and measurement probes more straightforward. While media can flow through channels in PDMS, diffusion through the material itself is limited. In order to overcome this diffusion limitation, chips have been designed in which the central area used for growing and observing cells is made from patterned agarose gel. The patterned agarose pad can then be sealed into a PDMS support, and fluid flow can be directed through the delicate and soft agarose structures. One example of this type of chip is that designed by Moffitt et al. (2012). More details on the methods used to construct such agarose and PDMS devices are given in Section 11.

3 FLUID FLOW

Accurate, predictable and reliable control of fluid flow is vital for the smooth running of a microfluidics system. There are a number of decisions that need to be made including what type of pump to use to drive the flow and how to transmit the fluid from the pump into the chip.

Three main types of pumps can be used to deliver media to microfluidics chips: peristaltic pumps, syringe pumps and pressure controllers. Peristaltic pumps are simple to use, but are less favoured for bacterial work, as they are not very accurate at low flow rates. Syringe pumps are relatively inexpensive and provide an easy to set up solution to deliver media. Multiple syringe pumps connected to different inlet ports can provide not just media but also antibiotics or other chemicals when gradients are required. The amount of media delivered to a microfluidics chip using a syringe pump is easily measured and predictable. The main drawback of using this type of pump is that they deliver media in a pulsed manner and are not very responsive to changes in flow rate. Depending upon the resistance in the tubing, it can take up to an hour to equilibrate the flow rate to the required velocity. This leads to uncertainty in the amount of fluid being delivered during the time it takes for the flow rate to equilibrate, with external flow metres being used to estimate the actual volumes and rates of medium delivery. Additionally, if the microfluidic channel attached to the syringe pump becomes blocked, there is no feedback mechanism, so pressure will build up inside the device, usually leading to leakage from the ports or failure of the plasma bonding connecting the PDMS to the underlying glass surface. Either of these problems renders the chip unusable. An alternative to syringe pumps is to use a pressure-driven controller. Though these controllers tend to be more expensive, they are much more responsive to changes in flow rate and provide a much smoother, pulseless flow of fluid into the chip. Elveflow Plug and Play microfluidics provides a summary of the types of controllers available.

Transmission of the fluid from the pump into the microfluidic chip can be a problem as leaking can occur. The main causes of leakage are poorly fitting ports, blockage of the microfluidics channels (often with PDMS or cells) or too much pressure being applied from the pumps. In order to prevent leakage and blockage, a number of practical steps that can be taken. First of all, when ports are punched into PDMS, care must be taken to ensure that the core of PDMS is removed from the resulting hole; otherwise, this can be washed into and block the channels. Removing the core is made easier by using a biopsy punch that is slightly smaller than the size required for the post, for example, using a 0.75-mm diameter biopsy punch for a 0.8-mm port post. Thorough cleaning of the PDMS part of the chip and also the glass surface before plasma bonding will also help to remove any loose fragments of PDMS from the surface.

The other main cause of blocking is cell clumping. At the design stage, careful thought should be given to the design of the inlet and outlet channels and the buffer reservoirs. These channels can be made wider, and the chambers that form the reservoirs for the inlet and outlet ports can be made wider and rounder (Fig. 3, right), rather than narrow with cross-shaped ends, for example (Fig. 3, left). This reduces the number of small features in the vicinity of the ports for cells to get trapped in. It is also important to make sure that cell cultures or samples to be put into the channels are well mixed and in a single-cell suspension on entering the chip.

Valves can be a useful addition to the device if it is important to the experiment to have areas of the chip that can be separated off, or to control the precise timing or location of different components. This requires the use of a layer of pressurised air to control the flow at the valves. The control layer usually sits above the fluid layer, when using an inverted microscope. Valves can be designed in different ways,

4 Additional considerations when designing microfluidics system

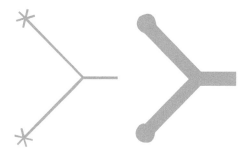

FIG. 3

Left, an example of a port and inlet that would be more prone to clogging. The centre of the star is the position of the inlet hole and port. *Right*, a simplified version of the port with wider channels, which is less likely to clog. Designs drawn in L-edit software.

but those shown by Grover, Ivester, Jensen, and Mathies (2006) and Unger, Chou, Thorsen, Scherer, and Quake (2000) are the most commonly used for devices made with PDMS.

4 ADDITIONAL CONSIDERATIONS WHEN DESIGNING AND SETTING UP A MICROFLUIDICS SYSTEM

In addition to the control of fluids, nutrients and chemical gradients, gas exchange and temperature in the device are also important considerations. Gas exchange is largely driven by the type of material used to produce the device. PDMS is permeable to gas; however, this permeability can be affected by surface modifications, for example, plasma activation or attachment of proteins or chemical species. The thickness of the PDMS can also affect gas diffusion, with Ochs, Kasuya, Pavesi, and Kamm (2014) showing higher oxygen levels in 2 mm thick PDMS chips than in those that were 5 mm thick. It is possible to overcome any oxygen limitation in the medium by keeping it well aerated and increasing the rate at which the medium is replenished in the system. However, if a super oxygenated environment is required, such as when simulating a highly aerated environment, or if an anaerobic environment is required, then modifications would need to be made to the chip design. A number of solutions to this have been devised. Lam, Kim, and Thorsen (2009) designed a differential oxygenator, allowing the dissolved oxygen levels within the microfluidics device to be controlled. This system uses oxygen and nitrogen gas mixed in a microchannel in close proximity to the channels containing the cells. With this device the authors were able to culture organisms with different oxygen requirements. In order to maintain anaerobic conditions, Steinhaus, Garcia, Shen, and Angenent (2007) produced a PDMS microfluidics chip using the soft lithography method. The device was then encased in a small acrylic chamber sealed with an O-ring and silicon grease. All tubing leading to the device was also encased in neoprene tubing.

The simplest method for maintaining a set temperature is to have an external incubation chamber that covers both the microfluidic device and most of the microscope. The volume of air to be warmed in such incubation chambers is, however, relatively large, so it does not respond quickly to changes in the required temperature. External incubation chambers also tend to be an expensive solution if not already available in the lab. Preheating the input liquids is also an option. Commercial systems tend to have built-in systems that facilitate the preheating of the media and also have some form of incubation chamber (e.g. CellASICs, BioFlux). The review by Miralles, Huerre, Malloggi, and Jullien (2013) on heating and temperature control in microfluidic systems provides an excellent overview of the relative strengths and weaknesses of the various heating methods.

5 COMPUTATIONAL ANALYSIS AND CONTROL OF BACTERIAL MICROFLUIDIC SYSTEMS

Computing hardware and software are an integral part of a microfluidics system. They are used in the design of the systems themselves, their spatiotemporal control and the analysis of the bacteria. In many cases, there is a requirement for the operating systems to be controlled in real-time.

The small size of bacteria in comparison to eukaryotic cells presents challenges, not only in relation to the scale of microfluidics fabrication but also for the computational analysis, particularly for image analysis. Examples of some useful programmes for image analysis are shown in Table 2. Tools such as MicrobeTracker (Sliusarenko et al., 2011) have proved valuable in this respect. MicrobeTracker allows outlines and segmented cells to be defined in images and minicolonies and also facilitates the tracking of cell lineage. An associated tool, SpotFinder, allows the foci of fluorescently labelled molecules inside bacterial cells to be tracked. Recently, MicrobeTracker and SpotFinder have been upgraded to Oufti (Paintdakhi et al., 2016). However, in many cases the image analysis techniques that have been developed for bacterial microscopy are not optimal for microfluidics-based analysis, and novel approaches have been developed that are tailored to these devices.

For example, super-resolution fluorescence microscopy methods can be used in combination with microfluidics systems to investigate the architecture, composition and dynamics of bacterial substructures. Cattoni et al. (2013) describe an automated cell detection and image analysis system for single-molecule localisation in cell populations, facilitating the study of heterogeneity dynamics and other properties at a super-resolution level. This system also incorporates features that allow cell-flatness, stability and growth in microfluidics chambers to be monitored as well as the detection and characterisation of cell clusters.

While microfluidic systems offer a powerful approach to the study of bacterial motility, suitable computer-based analytical systems are required to interpret the complex images that are generated (Son, Brumley, & Stocker, 2015). With the development of such systems, it is now possible to describe the mechanics of microbial

Table 2 Image Computational Analysis Tools for Microfluidic Microscopy

Programme	Source Code	How to Apply and Work	Last Updated Year	Reference or Contact	Website
Cell tracker	Matlab	Cell tracking and detection Statistical analysis of cell movement	2016	Piccinini, Kiss, and Horvath (2016)	http://www.celltracker.website/
Lineage mapper	Matlab	Overlap-based cell tracking and detection Assign tracks	2016	Chalfoun, Cardone, Dima, Allen, and Halter (2010)	https://isg.nist.gov/deepzoomweb/resources/csmet/pages/cell_tracking/cell_tracking.html
Manual tracking	ImageJ	Cell tracking	2005	Fabrice Cordelieres	https://imagej.nih.gov/ij/plugins/track.html
MicrobeTracker	Matlab	Outlines and segments of cells Cell lineage tracking	2012	Sliusarenko, Heinritz, Emonet, and Jacobs-Wagner (2011)	http://microbetracker.org/
MTrackJ	ImageJ	Cell and particle tracking	2015	Meijering, Dzyubachyk, and Smal (2012)	http://www.imagescience.org/meijering/software/mtrackj/
Object tracking software	Matlab	Cell tracking	2003	Nick Darnton	http://www.rowland.harvard.edu/labs/bacteria
ObjectJ	ImageJ	Length and diameter of cell Fluorescent spot detection Spore tracker	2016	Jake Jaff Norbert Vischer Stelian Nastase	https://sils.fnwi.uva.nl/bcb/objectj/

Continued

Table 2 Image Computational Analysis Tools for Microfluidic Microscopy—cont'd

Programme	Source Code	How to Apply and Work	Last Updated Year	Reference or Contact	Website
Oufti	Matlab/Stand-alone	Cell, spot and object detection Fluorescent signal outline and profile	2016	Paintdakhi et al. (2016)	http://oufti.org/
SpotFinder Z	Matlab	Fluorescent spot detection	2012	Sliusarenko et al. (2011)	http://microbetracker.org/
SpotFinder M Time Lapse Analyser	Matlab	Cell tracking and counting Proliferation and tube assay analysis	2009	Hans Armin Kestler	http://www.informatik.uni-ulm.de/ni/staff/HKestler/tla/
TLM-Tracker	Matlab/Stand-alone	Outlines and segments of cells Cell lineage analysis Quantification of signal detection	2012	Richard Münch Johannes Klein	http://tlmtracker.tu-bs.de/
TrackMate	ImageJ	Cell tracking Single particle tracking	2016	Jean-Yves Tinevez	http://Imagej.net/TrackMate

locomotion in different environments. These environments can include solid surfaces, flowing fluids and dense cells suspensions. For example, these analytical approaches can be used to define the 'hydrodynamic signatures' of swimming bacteria (Elgeti, Winkler, & Gompper, 2015). This signature represents the variation in the fluid flow around the microorganism, as defined by the motility strategy of the organism itself. Dynamic visualisation of motility and the associated hydrodynamic signatures is difficult but can be achieved using a high-speed imaging approach, combined with tracer particles in the medium surrounding the bacteria (Drescher, Dunkel, Cisneros, Ganguly, & Goldstein, 2011). This approach can be used to create cell trajectory visualisations that can indicate the overall swimming path of individual bacteria.

Tracking software has also been used to monitor the growth of bacterial cells grown in channels (such as the Mother Machine; Wang et al., 2010). In the authors' laboratory, we have developed a system that identifies and monitors single bacterial cells and uses a computational graph-based approach to describe the cell lineages and trajectories. The resulting data allow cell lineages to be analysed computationally using graph traversal algorithms or simply analysed visually (Fig. 4).

In addition to the analysis of cell images, computational approaches play an important part in the control of the environment within the microfluidic chips. Software can be programmed to control the flow of fluid in the system and to control the state of valves which can be used to sort bacterial systems. For example, Li et al. (2014) describe an innovative use of this approach to develop gradients of bacterial growth inhibitors and to automatically measure the response of the bacteria. The system can provide kinetic information on growth inhibition and the dynamics of any morphological changes in the bacteria over a range of chemical concentrations.

Increasingly, droplet-based microfluidics is becoming popular (Zhu & Fang, 2013). In these systems, bacteria are contained in liquid droplets that are suspended in oil. A variety of approaches are then used to characterise the droplets and their bacterial population. These include bright-field and fluorescence microscopy, electrochemistry, laser-induced fluorescence and Raman spectroscopy. One of the advantages of the droplet-based systems is that the computational analysis of the droplets can be used to make decisions about the manipulation of the droplet itself, thus combining analysis and control. The most widely used example of this dual functionality is in cell sorting where droplets containing bacterial cells expressing a particular phenotype are sorted in real time from the rest by physically manipulating the droplet itself (Huang et al., 2015).

6 MICROFLUIDICS FOR THE ENGINEERING OF BACTERIAL SYSTEMS

Microfluidics systems offer an attractive method for automating the process of engineering bacteria and also the process of analysing the resulting bacteria. Microfluidic devices have been described that automate the processes of bacterial DNA

84 CHAPTER 3 Microfluidics for bacterial imaging

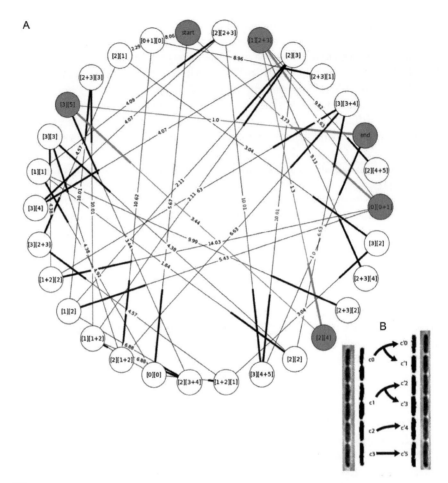

FIG. 4

Graph-based analysis of the growth of a single bacterial cell in a microfluidic channels. (A) Graph-based visualisations allow the lineages of cells to be identified; individual cells are nodes in the graph (*circles*) and the edge links (*lines*) indicate the division event leading to the daughter cells for each division. The starting mother cell is indicated at the top of the graph. Therefore, the mother and progeny cells can be defined by following the edges in the graph, either manual or through a software-based analysis. (B) A magnified view of part of the bacterial cell division events in 1 μm channels, that are used to generate the graph representation in (A). The lineage of the cells can be seen as the cells divide to produce daughter cells.

extraction (Matos et al., 2013), RNA extraction (Vulto et al., 2010) and the preparation of DNA libraries (Yehezkel et al., 2016). Droplet technology is proving especially useful in this respect and devices have been proposed to automate many of the steps in the field of synthetic biology (Shih et al., 2015).

The analysis of the phenotype and behaviour of recombinant bacterial systems can also be automated and carried out at a single-cell level (Guo, Rotem, Heyman, & Weitz, 2012). For example, microdroplet-based systems can be used in screening and sorting-based approaches to enrich for bacteria with a given phenotype from a mixed population. In some cases, bacteria have been encapsulated in agarose beads suspended in buffers to facilitate their sorting and manipulation. Eun, Utada, Copeland, Takeuchi, and Weibel (2011) utilised this approach to enrich for rifampicin-resistant mutants from a background nonresistant population. Microfluidic approaches also offer advantages for the study of quorum sensing in bacteria, both natural and synthetic (Fig. 5). The ability to monitor expression dynamics and cellular heterogeneity and stochasticity is also attractive. Ramalho and coworkers describe a microfluidics-based approach for the single-cell analysis of bacterial sender and receiver systems based on *Escherichia coli* engineered with quorum sensing systems derived from *Aliivibrio fischeri* (Ramalho et al., 2016). This approach allowed the dynamics of the gene expression response of receiver bacteria to varying amounts of the quorum sensing inducer N-3-oxo-C6-homoserine lactone (AHL) to

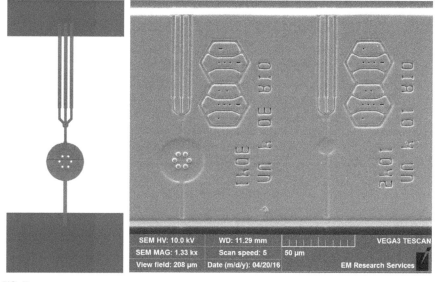

FIG. 5

Left, quorum sensing microfluidics design, produced at Newcastle University. Channels (the circular chamber and the three parallel channels) are designed to trap cells that act as senders and receivers of quorum sensing molecules. Quorum sensing molecules diffuse through the narrow diffusion channels between the cell chambers. The hexagonal structure contains notations about the location of the region of interest as the design is repeated hundreds of times on the same device. *Right*, an SEM image of the design after it has been fabricated using lithography. The device has been backfilled with an oxide material to achieve the final feature size, and then replica moulded from the wafer with PDMS.

be studied. Danino, Mondragón-Palomino, Tsimring, & Hasty (2010) demonstrated the use of microfluidic systems for investigating the collective synchronisation and spatiotemporal waves induced in bacterial populations by quorum sensing molecules (Danino et al., 2010). They were also able to engineer the genetic circuits to generate synchronised oscillations in growing populations of cells.

7 BIOFILM, MICROBIAL ECOLOGY AND SPECIES–SPECIES INTERACTIONS

Microfluidics has been widely used for studying single-cell dynamics in pure cultures. However, in order to understand how organisms function and respond to stresses and changing environmental conditions, systems for studying mixed communities need to be devised. In a review on current technologies for the analysis of bacterial microenvironments, Wessel et al. (2013) highlight the need for microbiologists to embrace new technologies and realise the limitations of pure culture techniques.

When researching complex bacterial interactions, such as the formation of multispecies biofilms, understanding spatial assembly rules governing how bacteria grow, survive and evolve is important. And an understanding of the relationships between host, biofilm and pathogen is vital for medical and engineering applications. Bacterial communities in nature, industry or medicine are seldom composed of bacteria of a single strain or species, and therefore, understanding the interplay between bacteria of different species is important. In a recent review, Burmølle, Ren, Bjarnsholt, and Sørensen (2014) posed the question 'Interactions in multispecies biofilms: do they actually matter?' After reviewing the literature in this area, the authors concluded that the interactions help to determine the function and development of biofilms and, as such, need to be taken into account for medical and industrial biofilm applications. They also highlighted a key challenge for the study of bacterial biofilms, and indeed all microbial systems, namely 'when diversity increases, so does complexity'. Laboratory methods traditionally used for pure culture-based research, including microfluidics, are now being adapted for assessing such interactions and complexity.

Microfluidics is particularly useful for studying bacterial interactions for many reasons. In microfluidic chips, small numbers of cells can be grown under very specific conditions, with tightly controlled and predictable fluid flow rates and nutrient conditions, including the ability to setup chemical gradients across the chip. Microfluidics also reduces the need for large quantities of growth media or other reagents, due to the small scale of the channels. This is particularly important where specific environmental fluids are used in place of traditional laboratory-produced growth medium, such as the use of saliva to assess the growth and effect of antimicrobials on dental plaque biofilms (Nance et al., 2013). The use of environmentally relevant fluids as sources of nutrients, and of bacterial cells to seed the community in such experiments, has the potential to allow growth in the laboratory of organisms that are unculturable in conventional laboratory media. Microfluidic chambers can also be designed with the specific requirements of multiple cell types in mind. For example, they have been used successfully for pathogen–host interaction studies, where the pathogens and their hosts are cultured in separate growth chambers and then exposed to each other in a controlled manner (Kim et al., 2010). The ability of microfluidic systems to be coupled with a variety of microscopy methods means that tracking of cell–cell interactions and biofilm formation in both time and space is achievable. The design of microfluidic systems and their small volumes has the added benefit of allowing for high throughput by simultaneously running multiple samples and controls.

7.1 MULTISPECIES BIOFILM MICROFLUIDICS

In this section, we focus on the latest microfluidic methods for growing and monitoring multispecies biofilms and how microfluidics is being used to determine spatial assembly rules, in both synthetic and natural bacterial communities.

Biofilm research is of great interest to industry. In many industries the build-up of biofilms is problematic, leading to increased costs and adverse health effects. Within the human body biofilms can cause tooth decay and gum disease (Marsh, 1994, 2005), and biofilms on medical devices, such as artificial joints and valves and catheters, can promote infection and affect patient recovery (Song et al., 2013). On the larger scale, biofilms in pipes, for example, those delivering drinking water, can harbour pathogens and potentially cause blockages. In the shipping industry, biofilms on the surfaces of hulls can increase drag and have a negative effect on efficiency, increasing fuel consumption and costs. On the other hand, in some industries, biofilms are required to fulfil a useful role. For example, in wastewater treatment trickling filters, the biofilm on a solid media (such as rock) reduces organic load and is responsible for treating the sewage. Similarly, the biofilms that form on the top layer of a sand filter are critical to the system's ability to purify drinking water.

Microfluidic systems have been used in many studies of biofilms, with the majority of these focusing on pure culture biofilm formation. Although these studies can be informative, they are clearly an oversimplification of what happens in the 'environment' outside the laboratory. There are a number of excellent reviews on the use of microfluidics systems for studying pure culture biofilms (Rusconi, Garren, & Stocker, 2014; Wessel et al., 2013).

A key decision already mentioned earlier is whether to use a commercially available microfluidic device or whether to design, manufacture and run your own system. Commercially available devices have been used to great effect for the study of multispecies biofilms. Indeed, the majority of multispecies biofilm microfluidics studies use these ready-made, all-in-one systems that integrate temperature control, microfluidics chambers and flow control into one easy-to-use system.

By far the most commonly used system for the study of biofilms is the BioFlux system. This system is commercially produced by Fluxion Biosciences and combines a fluid flow system with a 96-well plate style setup. The cartridges are available in a number of designs (typically a 48-well format), allowing for the simultaneous running of 24 samples or experimental conditions. Each of the two wells in a pair is joined by a microfluidics channel. Channels vary in size depending on the model, but typically have a depth of 70 μm. The wells are used as inlets and outlets for cells and media, as shown in Fig. 6.

Nance et al. (2013) use the BioFlux system along with confocal laser scanning microscopy to study the effects of antimicrobials on dental plaque biofilms. This novel approach, using bacteria from pooled saliva samples to seed the system and cell-free saliva as a growth media led to the growth of biofilms, more closely

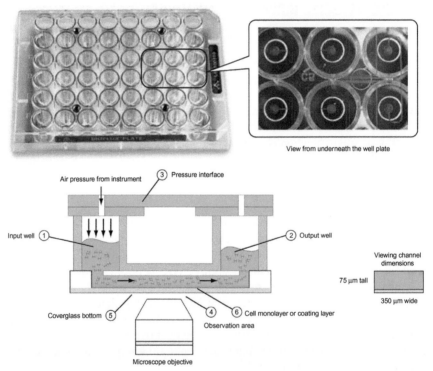

FIG. 6

BioFlux microfluidic device set up. *Top left*, plate format. *Top right*, detail of microfluidic channel between the wells, viewed from below. *Bottom*, schematic of the system. Cells and reagents are loaded into the input well (1), which is connected to the outlet well (2). The BioFlux pressure interfaced applies controlled pneumatic pressure across the top of the wells (3), to drive the flow rate. The sample (6) is observed (4), through the base of the plate (5), a 180-μm glass plate.

Reproduced with permission from Fluxion Biosciences (http://support.fluxionbio.com/hc/en-us/articles/203648088-BioFlux-Product-Family-Brochure).

resembling biofilm formation as it is thought to occur on teeth. These biofilms were then challenged with antimicrobial compounds at different concentrations and in situ live/dead staining was used to assess their efficacy. Biofilms were later flushed from the chip, using a high shear stress pulsing flow regime to dislodge the cells from the glass surface. The resulting biofilm was then the subject of bacterial tag-encoded FLX amplicon pyrosequencing (bTEFAP), to confirm that the biofilm that was formed resembles a natural tooth surface biofilm.

Following on from this work, Kolderman et al. (2015) also used the BioFlux system to grow oral biofilms from bacterial cells within pooled saliva samples. In this study the effect of the concentration of L-arginine HCl (LAHCl) on biofilm structure was assessed. This compound has been shown to destabilise biofilm aggregation at high concentrations and mediate cell–cell signalling at lower concentrations. The combination of LAHCl and a common antimicrobial compound found in oral health products, cethylpyridinium chloride (CPC), was tested for efficacy. CPC is known to be an effective antimicrobial, but has difficulty penetrating biofilm surfaces. The study therefore aimed to assess whether using the two treatments in combination would increase biofilm cell death and disaggregation. The ability to image microfluidic channels by confocal scanning laser microscopy means that a wealth of information could be assembled and analysed through image processing. In this study the authors calculated biovolume, average biofilm thickness, biofilm roughness and cell viability (based on live–dead stain fluorescence) using the COMSTAT and IMAGEJ software.

Another variation of the BioFlux system, the 'Invasion' plate, was used by Oliveira et al. (2015) to investigate the effects of low concentrations of antibiotics when strains of bacteria are combined. The Invasion plate has two inlets and one outlet per microfluidic channel. This allows media containing antibiotics to be added, with the other channel containing media only, resulting in the formation of an antibiotic concentration gradient. A range of concentrations could be tested in a single experiment and this, coupled with confocal laser scanning microscopy and image processing, allowed the cells' responses to be recorded directly and analysed. Attachment experiments using the microfluidics device suggested that the cells have an increased likelihood of attaching when antibiotics are present, indicating that biofilm formation was actually promoted by the presence of antibiotics. The authors highlight, however, that the 'direct assessments of cell division in three-dimensional biofilms is challenging' and was not done in the study. This is an important limitation of the current microfluidic methods available for analysing biofilm formation. It is challenging to form 'true' environmentally relevant biofilms in a microfluidics channel and to design a channel that is also optimal for imaging. Imaging is much simpler when the cells are effectively growing in a monolayer, as in the Mother Machine and the agarose pad microfluidics discussed in Section 8. Once complex three-dimensional structures begin to form, monitoring becomes much more difficult. Potentially, this could be resolved by advances in image processing techniques, a key area for the future development of the field.

7.2 SPATIAL ARRANGEMENTS AND INTERACTIONS

In addition to the study of multispecies biofilms, microfluidics devices have begun to be used to investigate the rules and interactions between bacteria and how this may affect spatial arrangements of bacteria in the environment and in industrial settings. PDMS microfluidics devices are well suited to studying spatial arrangements and interactions, as micron-sized chambers can be designed with varying degrees of connectivity and to hold cells in a range of formations.

By varying the distances between chambers containing three different species of bacteria, Kim et al. (2008) were able to build a stable synthetic community of wildtype soil bacteria with syntrophic interactions. This would not be possible in flask-scale studies, where synthetic communities tend to be unstable. The three bacterial species were specifically chosen because they do not compete for the same nutrients, but each produces compounds critical to the metabolism or survival of the others. The chambers were designed to allow the free flow of metabolites and other chemical cues into the main channel, while keeping the three species at a defined distance from one another. This kind of chip could therefore be useful for gaining a better understanding of communities of organisms, for example, where the biodegradation of pollutants is carried out in multiple stages by different species.

8 CELL CYCLE ANALYSIS AND SIZE HOMEOSTASIS STUDIES

Bacteria propagate by repeated cycles of growth and division, referred to as the cell cycle. During the progression of the bacterial cell cycle, from a new born cell to its division to become two new born daughter cells, three major events occur: DNA replication, chromosome segregation and cell division. Thus, while a new born cell grows in size, it duplicates its genome and segregates the sister chromosomes to appropriate locations in the cell, before making a division septum at the appropriate time and location to generate the two daughter cells. These events occur independently but are well coordinated to ensure the fitness of the progeny (Adams, Wu, & Errington, 2014; Wu & Errington, 2011).

Cell size control is a fundamental problem for all living organisms. Bacterial cells normally divide only when they are near double the new born sizes; therefore, their size distributions are relatively narrow. It is still not clear how the cell cycle is adjusted so that the cell division is coupled to growth under different growth conditions. For over 40 years, the 'sizer' model and the 'timer' model were accepted as the possible mechanisms for cell size control. The 'sizer' model proposed that bacteria cell size homeostasis was maintained by sensing cell size and that cell division was triggered when a critical size was reached (Donachie, 1968). The 'timer' model, on the other hand, suggested that the timing between DNA replication and cell division remained constant across different growth rates (Cooper & Helmstetter, 1968). Most of the experimental evidence for these models came from population-averaged data.

Studying size homeostasis requires precise measurement of cell size (length) of a large number of cells grown under steady-state growth conditions. This was achieved recently using microfluidic microscopy. Campos et al. (2014) and Taheri-Araghi, Bradde, et al. (2015) were able to maintain constant growth conditions over a long period of time, so that cells could be imaged and analysed over several generations. The techniques also allowed the authors to monitor and track many cells simultaneously, and to measure cell length from birth to division with high precision at the single-cell level. Both studies show that size homeostasis is maintained through an 'adder' rule, where cells sense how much they have grown, and divide when they have added a constant size (volume) since birth, irrespective of the birth size (Campos et al., 2014; Taheri-Araghi, Bradde, et al., 2015). The 'adder' rule has now been shown to apply to several evolutionarily divergent bacterial species such as *Caulobacter crescentus*, *E. coli*, *Bacillus subtilis* and *Pseudomonas aeruginosa* (Sauls, Li, & Jun, 2016). Taheri-Araghi, Bradde, et al. (2015) noted that a similar model, termed an 'incremental model', was proposed by Koppes and colleagues in the 1990s but was not well accepted. This was due in part to the lack of a 'comprehensive set of data containing distributions of both size and time between successive cellular divisions' (Taheri-Araghi, Bradde, et al., 2015; Voorn & Koppes, 1998; Voorn, Koppes, & Grover, 1993), which is exactly what microfluidic microscopy is capable of producing.

Although the microfluidic devices used by the two groups had different designs, both trapped cells in spaces of single layer cell height to ensure monolayer growth. The setups were able to support continuous medium flow, efficiently delivering fresh medium to the growth channels/chambers, at the same time removing cells that emerged from the growth chambers/channels (Campos et al., 2014; Taheri-Araghi, Bradde, et al., 2015; Ullman et al., 2013; Wang et al., 2010). The microfluidic continuous culture device used by Taheri-Araghi, Bradde, et al. (2015), also called the Mother Machine, was shown to deliver culture medium into the channels (by diffusion) much faster (~ 1 s) than the timescale of nutrient uptake (~ 2–3 min) (Wang et al., 2010).

The Mother Machine was developed by Jun and colleagues, and the technique has been made accessible through publications in various journals and websites from the Jun lab (Wang et al., 2010; Taheri-Araghi, Bradde, et al., 2015; http://jun.ucsd.edu/) (Method 1 in Section 11). The 'Machine' is made from PDMS and mounted on a glass slide or coverslip. Instead of having growth chambers like that used by Campos et al. (2014) and Ullman et al. (2013), the Mother Machine features individual growth channels similar to those used by Balaban et al. (2004). The microscale features (i.e. channels and trenches) are replicated onto PDMS from a silicon/SU8 photoresist master mould that contains the negative of the features. Each chip can contain over 1000 growth channels, each approximately 25 μm (L) × 1.5 μm (W) × 1.4 μm (D) though the size of the channels can be modified. One end of the channel is closed and the other end opens into a main trench where growth medium flows through (Fig. 7). The cell at the closed end of the growth channel, distal to the trench, is referred to as the 'mother cell'. The pole of the mother cell

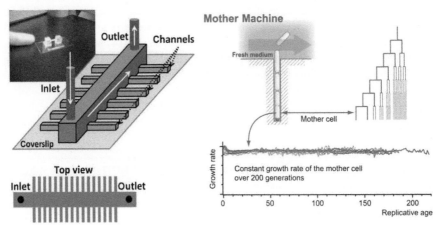

FIG. 7

The Mother Machine. The Mother Machine is made with PDMS from the master mould (*left*). The microfeatures on the master mould were made in two steps, the first layer generates the growth channels and the second layer forms the trench. The growth channels are connected to a main trench of varying lengths and widths. Growth medium flows into the device from the inlet, delivering fresh nutrients into the growth channels and ensuring steady-state conditions for all cells (*right*). Cells emerging from the channels are removed by the medium flow into the main trench. The machine allows thousands of cells to be imaged for hundreds of generations.

From fig. 1 and Supplemental Figure—Design of the Mother Machine of Wang, P., Robert, L., Pelletier, J., Dang, W. L., Taddei, F., Wright, A., & Jun, S. (2010). Robust growth of Escherichia coli. Current Biology, 20, 1099–1103.

that abuts the closed end of the channel is inherited from one generation to the next. The size of the growth channels is designed to match the average diameter of the bacteria to prevent the mother cell from moving around, making image analysis and cell tracking much easier. The Mother Machine is relatively easy to make and use, however, it requires a custom-made master mould (http://jun.ucsd.edu/) to produce the PDMS chips, and different moulds are likely to be required for different bacteria or, even different growth conditions under which cell width is expected to change significantly.

To maximise the potential of such devices, several groups have developed microfluidic systems using chips made of agarose instead of PDMS (Ducret et al., 2009; Moffitt et al., 2012; Wong et al., 2010). Various bacteria have been shown to achieve normal growth on agarose (Bergmiller, Pena-Miller, Boehm, & Ackermann, 2011; de Jong, Beilharz, Kuipers, & Veening, 2011; Joyce, Robertson, & Williams, 2011; Lindner, Madden, Demarez, Stewart, & Taddei, 2008; Stewart, Madden, Paul, & Taddei, 2005; Young et al., 2012) and, as a result, agarose has been widely used as a solid support for bacterial growth during microscopy. One of the advantages of agarose over PDMS is its softness, which means that it is less likely to exert mechanical stress on cells (Mannik, Driessen, Galajda, Keymer, & Dekker, 2009; Moffitt et al., 2012). Agarose is also more permeable to aqueous solutions than PDMS (Moffitt et al., 2012; Qin, Xia, & Whitesides, 2010), an important consideration for experiments that require rapid medium diffusion and small molecule exchange. It has also been reported that uncured PDMS can leach and accumulate in cells, though the effects have not been fully investigated (Moffitt et al., 2012; Regehr et al., 2009).

The agarose-based device designed by the Cluzel lab, called 'the single-cell chemostat', has growth channels printed on agarose rather than on PDMS, and the channels open at both ends into gutters or trenches (Moffitt et al., 2012; http://labs.mcb.harvard.edu/Cluzel/; Fig. 8). The authors (Moffitt et al., 2012) showed that the

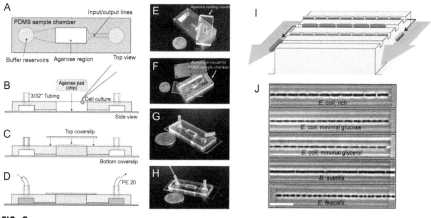

FIG. 8

The single-cell chemostat of Moffitt et al. (2012). (A) Top view of the PDMS sample chamber. (B) Side view of a partially assembled chamber. The bottom cover glass and the Tygon tubing have been added to create a chamber ready for cells and the printed agarose pad. The chamber is connected to the reservoirs at the bottom. Cells are added to the chamber and confined between a patterned agarose pad and a glass coverslip (the bottom coverslip). (C) Once the cells and the pad are loaded, an additional cover glass is laid over the agarose pad, and compresses and seals the device. (D) PE tubing is then used to introduce and remove buffer. (E) A PDMS mould for casting featured agarose pads, and aluminium mould used to fabricate it. (F) A PMDS sample chamber with the aluminium mould used to create it. (G) A partially assembled chamber with Tygon tubing. The PMDS sample chamber is now bonded to a bottle coverslip. (H) A fully assembled device with top coverslip and the agarose chip. (I) Medium flow pass both ends of the channels and removes cells as they emerge from the channels. (J) Soft agarose channels accommodate a range of bacterial morphologies.

Reproduced from Moffitt, J. R., Lee, J. B., & Cluzel, P. (2012). The single-cell chemostat: An agarose-based, microfluidic device for high-throughput, single-cell studies of bacteria and bacterial communities. Lab on a Chip, 12(8), 1487–1494 (figs. 3 and S3) with permission from The Royal Society of Chemistry.

agarose channels of this device were able to accommodate a range of bacterial morphologies and, to a degree, different sizes, without perturbing cell morphology or growth. Furthermore, by adjusting the flow rate of the growth medium, this device was also suitable for studying proximity-dependent growth in mixed microbial communities. To generate features on agarose surfaces, the authors fabricated a silicon master containing the positive of the features, which were replicated onto a PDMS intermediate mould with the features in negative. The features were then transferred onto an agarose pad from the PDMS intermediate. This potentially reduces the use of an expensive silicon master and prolongs its life. In addition to a master mould bearing the microfeatures, to make 'single-cell chemostats' two further matching custom-made aluminium moulds are required: one for making the PDMS sample chamber that will house the agarose pad and also form medium/buffer reservoirs; the other for making the PDMS mould used for casting patterned agarose pads (chips). These two matching moulds were designed such that the cavity in the PDMS sample chamber is similar in size to the agarose pads but slightly undersized ($\sim$5%) in both width and height. The compression needed to fit the agarose pad into the undersized chamber insures that cells are held firmly in place on the bottom cover glass, and that buffer must flow through gutters and not around the pad (Method 2 in Section 11; Moffitt et al., 2012).

9 CELL SHAPE AND GEOMETRY STUDY

Bacteria come in a variety of different shapes and sizes. However, each genus has its characteristic shape. Interestingly, although some bacterial species do go through morphological changes, this often occurs as a response to environmental cues. It is therefore believed that the shape of a bacterium has biological relevance (Young, 2006, 2007). For bacteria that have a cell wall, their shapes are defined by the cell wall, which is composed mainly of peptidoglycans (Cabeen & Jacobs-Wagner, 2005; Errington, 2015). Peptidoglycan is a macromolecule with long, rigid glycan strands covalently linked through short and flexible peptide bridges (Silhavy, Kahne, & Walker, 2010; Typas, Banzhaf, Gross, & Vollmer, 2012). Many genes involved in the synthesis of peptidoglycan and in bacterial shape determination have been identified, and the genetics of bacterial morphology has also been well studied for several model bacteria. However, the molecular mechanisms of bacterial morphogenesis and cell shape maintenance are still unknown. It is understood that the cell shape is the product of a combination of physical and biochemical processes, but due to technical restrictions it is difficult to study how shape-determining proteins exert controls on cell geometry. Similarly, it is also not clear how a cell regulates its shape in response to external mechanical stress. The advances in microfluidic microscopy have made it possible to address such questions.

The bacterial cell wall needs to have high strength and be sufficiently rigid to resist the internal turgor pressure, yet be dynamic to allow cell growth. The model Gram-negative bacterium *E. coli* and Gram-positive *B. subtilis* are both rod-shaped, which is the most common bacterial shape (Errington, 2015). While some bacteria grow by inserting new cell wall material only at the tips or at the division site, *E. coli* and *B. subtilis*, like many other bacteria, grow mainly in length, by synthesising new cell wall at multiple sites in the lateral wall of the cell ('dispersed' elongation) (Daniel & Errington, 2003; Silhavy et al., 2010; Typas et al., 2012). Back in 2005, the Whitesides lab fabricated shaped agarose microchambers and used them to generate filamentous (nondividing) *E. coli* cells of various shapes: crescent, zigzag, sinusoid and spiral (Takeuchi, DiLuzio, Weibel, & Whitesides, 2005). The device also allowed the cells to be released from the chambers into solution. The authors showed that upon being released from the microchambers and grown in solution, the cells retained their shapes for as long as they were prevented from dividing, but reverted back to rod shape when allowed to divide, demonstrating that bacteria could readily deform in response to external mechanical forces but were able to recover their shapes. More recently, several groups have used microfluidic microscopy to further study the effects of external mechanical stress on cell shape in *E. coli* and *B. subtilis* (Amir, Babaeipour, McIntosh, Nelson, & Jun, 2014; Amir & van Teeffelen, 2014; Caspi, 2014; Mannik et al., 2009; Si, Li, Margolin, & Sun, 2015). Using purpose-designed microfluidics chips that imposed geometric confinement or hydrodynamic forces in microfluidic devices, these studies demonstrated both the elastic (reversible) and plastic (irreversible) nature of the growing bacterial cell wall. Interestingly, Mannik et al. (2009) also found that *E. coli* cells were able to penetrate and grow in spaces much smaller than their diameters and readily adapt to the given shapes, even growing as flat, pancake-like cells, whereas *B. subtilis* cells were not able to grow in spaces that were smaller than the diameter of the cell. The lower morphological plasticity of *B. subtilis* is probably due to the difference in cell wall composition between Gram-negative and positive bacteria. Recently, Wu, van Schie, Keymer, and Dekker (2015) also used shaped microchambers (Fig. 9), in combination with drugs such as Cephalexin and A22 that interfered with cell wall synthesis, to sculpt *E. coli* cells into various shapes, and show that Min proteins,

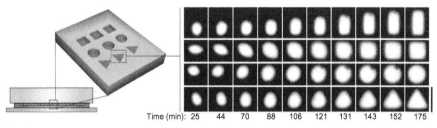

FIG. 9

The 'cell-sculpting' technique of the Dekker lab (Wu et al., 2015). *Left*: schematic of the cell-sculpting device, composed of a microscope coverslip (*bottom*), PDMS microchambers (*middle*, feature side up) and an agarose pad supplemented with nutrient and drugs (*top*). *Right*: cytosolic eqFP670 fluorescence (near-infrared fluorescence with emission maximum at 670 nm) images of *E. coli* cells growing into defined shapes.

From fig. 1 of Wu, F., van Schie, B. G., Keymer, J. E., & Dekker, C. (2015). Symmetry and scale orient Min protein patterns in shaped bacterial sculptures. Nature Nanotechnology, 10, *719–726.*

involved in determining the site of cell division, could sense and adapt to the morphological features of cells. Like most of the microfluidic chips, these microfeatures were printed from a silicon master mould onto PDMS but while doing so, the authors backed the PDMS film with a coverslip. After loading the bacteria into the microchambers on the PDMS chip, the cells were covered with an agarose pad containing growth medium. The device was then covered with wet tissues and parafilm to prevent drying while allowing air exchange. Although this microfluidic setup did not have continuous medium supply, the device itself is relatively simple to produce and assemble, and would be suitable for experiments that do not require long-term growth.

10 FUTURE PERSPECTIVES

Microfluidic technology has greatly empowered microscopy and allowed us to carry out research that was difficult or impossible to do before. However, with the possibilities come challenges. At present, commercially available microfluidic systems are only suitable for certain types of experiments. From the studies showcased earlier, it is clear that different research needs require chips of different designs. Therefore the application of microfluidics to bacterial imaging is limited by the design and by the availability of microfluidic chips. However, companies are now emerging that provide services for fabricating custom-made silicon moulds and even bench-top fabrication centres. Additionally, new tools are being created for automated design, fabrication and fluid flow control in microfluidics, making microfluidic technology more accessible for laboratories that do not have clean rooms and fabrication facilities.

The recent studies using microfluidic microscopy have provided rare insights into the behaviours of bacteria. However, many fundamental questions remain to be addressed. For example, the molecular mechanism of the 'adder' rule that controls bacterial size homeostasis is still unknown; we now know that bacteria adapt their shapes to external forces and constraints but we do not know how the shape change is brought about: is it by altering the distribution or by affecting the function of a protein or proteins? It is also not known how internal mechanical stress influences cell morphology. Microfluidics systems are only just beginning to be explored as methods for the study of multispecies bacterial communities and ecology. This is an area for which rapid prototyping of microfluidic chips to test multiple designs in a short space of time would be of particular benefit.

While the availability of custom-made microfluidic moulds and chips will inevitably widen the application of microfluidics in bacterial imaging, one possible concern is the reproducibility of the experimental conditions by other laboratories. It is therefore important that biologists and micromechanical engineers collaborate closely to create stable and highly definable devices. Similarly, the microenvironments in microfluidic devices can be difficult to characterise, so we also need the expertise of physicists and chemists to help generate devices with optimal medium flow and molecule exchange. Finally, it is also important that microfluidic devices are designed to be easy-to-use and are applicable to standard microscopy so more research can benefit from the power of microfluidic microscopy.

11 METHODS

As a general consideration, gloves should be used at all time when preparing and handling microfluidic devices.

Method 1 PDMS-based Mother Machine microfluidic device (Adapted from 'The mother machine handbook' of Jun Lab, University of California, San Diego, USA; http://jun.ucsd.edu/; Taheri-Araghi, Bradde, et al., 2015; Wang et al., 2010)

    A. Materials
1. Sylgard 184 Silicone Elastomer kit (containing silicone elastomer base and curing agent).
2. Pentane.
3. Acetone.
4. 50 mg/mL BSA.
5. Growth medium.

    B. Equipment
1. Vacuum chamber.
2. Harrick Plasma system.
3. Oven or hot plate for 65°C incubation.
4. Fume hood.
5. Tubing for medium and waste transfer (e.g. 0.023″ inner diameter).
6. Needles with Luer-lock interface (22 gauge).
7. Large syringe (e.g. 60 mL) with a Luer-lock tip for the medium.
8. Scalpel.
9. Syringe pump.
10. Magnetic stirrer.
11. Tweezers.
12. Harris Uni-core 0.75 mm punch.
13. Plastic Petri dishes.

    C. Preparing supporting PDMS blocks to bolster the PDMS chip.
1. Mix 42 g of PDMS with 4.2 g of curing agent (10:1 ratio) in a standard sized Petri dish. Stir for 30 min and degas using a vacuum chamber until the mixture is clear and bubble-free. During degassing, periodically disturb the vacuum with a brief influx of air to beak the bubbles and reduce spillage.
2. Cover the dish and place in a 65°C oven to cure for at least 2 days.

D. Printing micropatterns from the silicon/SU-8 wafer mould onto PDMS
  1. Line a large Petri dish with aluminium foil, making sure the foil is flat and smooth. Place the wafer mould in the centre of the dish, feature side up. Blow off any dust then replace the cover to keep clean.
  2. For a wafer of 4″, mix 9.0 g Silicone elastomer base with 0.9 g curing agent (10:1 to a total of 10 g) for 10 min until the mixture looks milky and bubbly, then degas using a vacuum chamber.
  3. Pour the degassed elastomer base/curing agent mixture onto the wafer, tilt the Petri dish to help spread the mixture and cover the entire wafer without spilling.
  4. Cover the Petri dish and place at 65°C to cure PDMS overnight.
  5. To peel the cured PDMS film off the wafer, first use the tip of a sharp scalpel to separate the two at the edge, then use a pair of tweezers to slowly and carefully peel the PDMS film off the wafer. The peeling motion should be done along the axes of the growth channels. Place the PDMS film in a clean Petri dish.
  6. Gently wipe off any uncured PDMS from the back of the wafer with a fibre-free wipe. Do not wipe the feature side. Store the cleaned wafer covered in a clean dish.
  7. Cut out the microfluidic chip with a scalpel, leaving about 1 cm along each side of the large channels, and about 5 mm around any fine features. Punch a 2 mm hole at one corner, away from the features, so it can be hung up later.
  8. Chemically treat the PDMS chip to remove any residual uncured PDMS. This procedure should be carried out in a fume hood. Wash the PDMS chip firstly in pentane for 2 h with stirring, then transfer quickly to acetone and wash three times, 2 h each wash. 200 mL of pentane/acetone in a 500-mL beaker is sufficient to treat 20 chips. Set the stirring speed to one to five rotations per second and cover the beaker during washing to prevent solvent evaporation. PDMS chips can be picked up with tweezers, and when dropped into the liquid, care should be taken to avoid partially submerging the chip, which could cause deformation.
  9. Hang the PDMS chip in fume hood to dry overnight.
E. Assembling the microfluidic device
  1. Identify the feature side of the PDMS chip. When looking at the PDMS chip, if the letters and numbers on the chip are readable, then you are looking at the nonfeature side.
  2. Place the PDMS chip, feature-side down, in a clean Petri dish. Cut out thick PDMS blocks (from A) and place them in the dish along with the chip, smoothest side up. Then bolster the inlet and outlet areas of the PDMS chip with the PDMS blocks using a plasma cleaner as follows:

First, place the dish into the plasma machine and expose the cleaned surfaces of the PDMS chip and the PDMS blocks to oxygen plasma according to the instruction for your machine (e.g. 15 s at 30 W in a Harrick Plasma system), then immediately lay the blocks onto the PDMS chip (exposed sides facing each other), covering the inlet/outlet areas, and press firmly to expel any air and to ensure complete contact.

Place at 65°C to bond for 10 min. Oxygen plasma makes exposed PDMS and glass reactive, so that covalent bonds form between surfaces brought into contact with one another. The PDMS block should be large enough to cover all fluid inlets and outlets of the chip. Dust and dirt on PDMS surfaces can be removed using scotch tape before plasma cleaning.

3. Locate the trench and punch one hole at each end, to create an inlet and outlet for the fluidic tubing, making sure that the hole goes all the way down through the PDMS chip/PDMS block. Each device can have more than one inlet/outlet. Ensure that the coring tool matches the size of the tubing to be used for supplying culture media (e.g. 0.75 mm coring tool for tubing of 0.023" inner diameter).

4. Seal the PDMS chip onto a glass bottom dish (e.g. Willco Wells or ibidi μ-Dishes) with no.1.5 glass thickness (170 μm) or a no.1.5 glass cover slip, again using a plasma cleaner (e.g. exposing the surfaces to oxygen plasma for 15 s followed by bonding at 65°C for 10 min). The glass surface should be cleaned beforehand by wiping with ethanol and then immediately with water to remove residual ethanol. When laying the chip down onto the glass, start at one end then gently tap away bubbles with blunt tweezers.

5. Bake the plasma bonded PDMS and cover slip or glass bottom dish to set for 10 min.

6. Inject 10 μL 50 mg/mL BSA into the device inlet using a standard 10 μL pipette tip. To avoid introducing air bubbles when injecting the solution, plunge the pipette down to the first stop and hold until the fluid flushes through and emerges from the outlet, then carefully withdraw the pipette tip from the inlet with the pipette still plunged. Seal the inlets and outlets with tape. Incubate the device, uncovered, at 37°C for at least 1 h to passivate the channel walls: the passivation solution will be drawn into the channels via evaporation. The microfluidic device is now ready for loading with the test bacteria. Note that if the device shows any evidence of leakage, it should be discarded.

F. Inoculating bacteria into the microfluidic device
   1. Concentrate the bacterial culture by centrifugation to, for example, an OD_{600nm} of 4–10. Use a 10 μL pipette tip to inject 10 μL of

concentrated culture into the device inlet, again plunging the pipette down to the first stop and holding until the bacterial suspension reaches the growth channels. Carefully remove the pipette tip as before and seal the inlet and outlet with tape. Incubate the loaded device uncovered, at the desired temperature to allow cells to be drawn into the growth channels. Check the device under microscope to monitor loading.

2. Prepare tubing for transferring culture media. Polyethylene tubing (e.g. with a 0.023″ inner diameter) can be used. Insert a blunt ended needle into one end of the tubing, then bend the needle to a right angle and twist off the Luer-lock hub. The ~1/2″ length of metal after the bend will later be inserted into the inlet hole. Insert another blunt ended needle into the tubing and connect it to the medium syringe through the Luer-lock hub.

3. A syringe with a Luer-lock tip can be used to house the growth medium. Install the syringe, containing the medium and connected to the tubing, onto a 'syringe pump'. Pump the medium through the tubing until all the air in the tubing has been displaced (a pump rate of 10–20 mL/h can be used for this step), then remove the tape from the inlet on the PDMS block and insert the bent metal needle end of the tubing into the inlet hole. There should be a drop of medium suspending from the end of the tubing as it is inserted into the inlet hole to prevent any bubble from entering the system.

4. Pump the medium through the device to flush all cells out of the main trench, until the fluid emerging from the outlet appears clear. A flow rate of 5–20 mL/h can be used for this step. Beware that higher rates can generate excess pressure and cause leaks.

5. Turn the pump rate down to the speed appropriate for the experiment (e.g. 1 mL/h). Insert a tube into the outlet hole to collect the waste into a bottle.

6. Mount the device on the microscope stage and begin imaging. Check for leaks continually.

Notes:

1. Add BSA (0.5 mg/mL) to the growth medium.
2. Mother Machine devices need to be optimized for different experiments and different bacteria, e.g., some require larger growth channels.
3. Videos for reference are available at: http://www.youtube.com/watch?v=RGfb9XU5Oow; http://www.youtube.com/user/JunLabHarvard?feature=mhum

Method 2 Agarose-based microfluidic devices (single-cell chemostats) (Fig. 8; adapted from Moffitt et al., 2012; http://dx.doi.org/10.1039/C2LC00009A; http://labs.mcb.harvard.edu/Cluzel/)

A. Materials
1. Tridecafluoro-1,1,2,2-tetrahydrooctyl-1-trichlorosilane
2. Copolymer 1: (7–8% vinylmethylsiloxane)–(dimethylsiloxane) copolymer.
3. Copolymer 2: (25–30% methylhydrosiloxane)–dimethylsiloxane copolymer, hydride terminated, 30–50 cst.
4. Modulator: 2,4,6,8-Tetramethyltetravinylcyclotetrasiloxane.
5. Catalyst: Platinum divinyl-tetramethyl-disiloxane; Gelest.
6. Solvent: Hexane.
7. Isopropanol.
8. Acetone.
9. Low melting point agarose.
10. Sylgard 184 Silicone Elastomer kit (containing silicone elastomer base and curing agent).
11. Growth medium.

B. Equipment
1. Vacuum chamber.
2. Harrick Plasma system.
3. Oven or hot plate for 65°C incubation.
4. Fume hood.
5. Spin Coater.
6. Tubing for medium and waste transfer (e.g. 0.023″ inner diameter).
7. 30G syringe Luer-lock needle.
8. Syringe with a Luer-lock tip for the medium.
9. Scalpel.
10. Syringe pump.
11. Magnetic stirrer.
12. Tweezers.
13. Biopsy punch (0.75 mm).
14. Glass Petri dishes.
15. Glass coverslips 22×50 mm and 18×18 mm.
16. Agarose casting mould, made of PDMS from an aluminium mould custom-designed by the Cluzel lab.
17. Aluminium mould for making PDMS sample chambers, custom-designed by the Cluzel lab, and a glass spacer that matches the mould.

C. Casting PDMS sample chamber.
1. Clean the aluminium mould with ethanol and tape around the edge of the mould to form a tray. Use multiple layers of the tape so the 'tray' has four reasonably rigid sides. Make sure that there are no points at which PDMS could leak out. Place in a glass Petri dish.

2. Mix 13.5 g of Sylgard 184 base with 1.5–1.7 g of the curing agent (8:1–9:1 ratio) in a large plastic weighing dish. Degas for 30 min in a vacuum chamber.
3. Pour about half of the degassed PDMS mixture into the mould. Use a sterile wooden stick to pop any bubble or move them away from the features.
4. Carefully lower the glass spacer into PDMS, starting at one end and working slowly along to prevent trapping bubbles underneath. Ensure that the spacer is pushed down, resting on the surface of the mould but not touching the round upright features that will form the inlet/outlet ports.
5. Pour in the remaining PDMS mixture. Place the mould onto the hot plate and leave to cure at 60°C in a fume hood for 24 h.
6. Remove the tape from the edges of the mould.
7. To remove the glass spacer, use a scalpel to cut down through PDMS and onto the glass, all the way along both edges of the spacer franked by the round upright features. Cut horizontally through the other two ends of PDMS so that the glass spacer can be released.
8. Carefully detach the PDMS block from the metal mould, working all the way around to prevent breakage.
9. Trim the short thick ends and then the long edges until it is slightly smaller than the glass coverslip that the PDMS chamber will be bonded to.
10. Use a scalpel to remove any excess PDMS from the bottom of the well where the agarose pad will fit. Make sure that the edges are smooth and there are no pieces of PDMS on the surfaces as this will affect bonding and placement of the agarose pad.
11. Use a biopsy punch to make holes in the centre of the round features for inlet and outlet ports. Push the punch right though and make sure to remove the sliver of PDMS that is pushed out of the hole. Ensure that the coring tool match the size of the tubing to be used.

D. Preparing the PDMS intermediate, consisting of a thin layer of hard PDMS (h-PDMS) bearing the features, supported by a soft PDMS backing of 3–5 mm.
1. Clean and silanize the master mould wafer by first rinsing the wafer with isopropanol, acetone and isopropanol. Dry with nitrogen and place the wafer on the cap of a small culture tube inside a vacuum chamber in a fume hood. This will allow both sides of the wafer to be salinized, and so prevent PDMS from sticking to the bottom of the wafer. Pipette 20 µL tridecafluoro-1,1,2,2-tetrahydrooctyl-1-trichlorosilane onto a fibre-less wipe

placed in a plastic Petri dish then place the dish in the vacuum chamber. Apply a vacuum for 30 min. Reintroduce air to the chamber very slowly to prevent the wafer from being thrown against the chamber.
2. To make h-PDMS, mix 13.6 g 8% vinylmethylsiloxane–dimethylsiloxane, 72 µL platinum divinyl-tetramethyl-disiloxane. To the mixture add 0.4 g 2,4,6,8-tetramethyl-tetravinyl-cyclotetrasiloxane, 4 g 25–30% methylhydrosiloxane–dimethylsiloxane and 2 g of hexane. Mix well then degas for 5 min.
3. Line the spin coater with aluminium foil then mount the wafer on the spin coater. Set the spin coater to spin at 100 rpm for 10 s, and start the machine. Slowly pour the mixture onto the spinning wafer, starting from the centre and move outwards until the wafer is completely covered. Carefully remove the wafer and place on a paper tower in the vacuum chamber. Degas for 3 min. Wipe off any spilt PDMS from the bottom of the wafer as it will interfere with the vacuum seal needed by the spin coater. Return the wafer to the spin coater and spin at 500 rpm for 5 s followed by 1000 rpm for 40 s.
4. Place the wafer on a wafer holder lined with aluminium foil (dull side up). Bake at 60°C in the oven for 1 h.
5. To make the soft PDMS (s-PDMS) backing, mix 70 g of Sylgard 184 base with 7 g of the curing agent (10:1 ratio) in a large plastic weigh dish. Degas for 30 min in a vacuum chamber. Gently pour the s-PDMS onto the wafer (on the wafer holder).
6. Allow the s-PDMS to settle and coat the wafer (~5 min), then place the wafer holder on a 65°C hot plate, cover and leave to cure at 60°C overnight to 24 h.
7. Carefully flip over the PDMS/wafer/aluminium foil and slowly remove the foil. Run a scalpel along the edge of the wafer to remove any PDMS that has leaked under the wafer.
8. Slowly and carefully peel the PDMS intermediate from the wafer. Clean the wafer and store covered. Cut out the PDMS intermediate.
9. Clean the PDMS intermediate before each use by rinsing in double-distilled (dd)H_2O, isopropanol and acetone, followed by isopropanol. Leave to air dry completely.

E. Printing patterned agarose pad from the PDMS intermediate.
1. Agarose pads are cast in an 'agarose casting mould' made of PDMS, onto which the PDMS intermediate is laid (from above) to replicate the microfeatures. 'Agarose casting moulds' are made of PDMS and fabricated using a custom-made aluminium mould,

designed by the Cluzel lab to match the mould for making PDMS sample chambers.
2. Melt low melting point agarose (3–5%) in growth medium by heating the mixture at 80°C until completely melted (about 30 min); about 130 μL agarose is needed for each pad. Vortex the agarose/medium mixture frequently to ensure that agarose has melted completely (several rounds of vortexing are needed, returning each time to the 80°C water bath). Leave the mixture for about 15 min after the last round of vortexing to allow bubbles to escape.
3. Place the cleaned PDMS intermediate on a cold block, feature side up.
4. In a 30°C warm room, pipette about 65 μL molten agarose onto the PDMS intermediate and another 65 μL into the agarose casting mould. Place the two together and press gently.
5. Leave to set for 1 h at 30°C.

F. Assembling the device and loading bacteria.
1. Bond the PDMS sample chamber from C11 onto a large glass coverslip (22 × 50 mm) using a plasma cleaner (18 W, 20 s, 1000 mTor atmosphere). Bake at 60°C for 10 min to strengthen glass-PDMS bond.
2. Cut two pieces of Tygon tubing, about 1/4" long and insert them into inlet and outlet. Seal any gaps between the tubing and the surrounding PDMS wall by adding a small drop of PDMS on the inlet and outlet, then baking at 60°C for 30 min.
3. Plasma clean the PDMS chamber from Step F1, and an 18 × 18 mm coverslip.
4. Add 1 μL of bacteria directly onto the glass surface in the PDMS chamber.
5. Gently separate the PDMS intermediate mould from the agarose pad. Carefully lift out the pad with a clean scalpel and place in the chamber, feature side down. The pad should go in one end first, then gently lower the other end to direct liquid to fill the gutters and push out air bubbles.
6. If any liquid pools on the top of the gel, remove it gently with a Kim wipe.
7. Place the plasma-cleaned coverslip over the pad and press gently to seal the PDMS agarose chamber.
8. Allow plasma bond to set for 15 min (the device can be placed at a temperature appropriate for your bacteria) before introducing growth medium into the device.
9. Place the PDMS intermediate mould in ddH$_2$O at 80°C for cleaning.

G. Connecting the device to growth medium
1. Connect the syringe containing the growth medium to tubing and install the syringe on a syringe pump. Load the tubing with medium.
2. Preload buffer reservoirs with medium. This is done by inserting a 30 gauge needle first into the rear buffer reservoir and connecting the tubing to the needle. Run the pump at 30 μL/min until the reservoir is full. Repeat for the front buffer reservoir.
3. Mount the device on the microscope stage. Run pump at 10–30 μL/min.

Notes:
1. PDMS intermediates can be made with a single layer of s-PDMS if the smallest features (i.e. tracks along which bacteria grow) are greater than 0.5 μm.

ACKNOWLEDGEMENTS

L.E. is funded by an EPSRC Frontiers grant (EP/K039083/1) awarded to A.W., S.L. and L.J.W. are funded by a Wellcome Trust Senior Investigator Award (WT098374AIA) to Jeff Errington and thank him for his support. We are grateful to Suckjoon Jun, Jeffrey Lee and Philippe Cluzel for providing and sharing information on their microfluidic devices, and to Enrique Balleza of the Cluzel Lab for sharing their protocols.

REFERENCES

Adams, D. W., Wu, L. J., & Errington, J. (2014). Cell cycle regulation by the bacterial nucleoid. *Current Opinion in Microbiology, 22,* 94–101.

Ahmed, T., Shimizu, T. S., & Stocker, R. (2010). Microfluidics for bacterial chemotaxis. *Integrative Biology, 2*(11–12), 604–629.

Amir, A., Babaeipour, F., McIntosh, D. B., Nelson, D. R., & Jun, S. (2014). Bending forces plastically deform growing bacterial cell walls. *Proceedings of the National Academy of Sciences of the United States of America, 111,* 5778–5783.

Amir, A., & van Teeffelen, S. (2014). Getting into shape: How do rod-like bacteria control their geometry? *Systems and Synthetic Biology, 8,* 227–235.

Balaban, N. Q., Merrin, J., Chait, R., Kowalik, L., & Leibler, S. (2004). Bacterial persistence as a phenotypic switch. *Science, 305,* 1622–1625.

Balagaddé, F. K., You, L., Hansen, C. L., Arnold, F. H., & Quake, S. R. (2005). Long-term monitoring of bacteria undergoing programmed population control in a microchemostat. *Science, 309,* 137–140.

Beebe, D. J., Mensing, G. A., & Walker, G. M. (2002). Physics and applications of microfluidics in biology. *Annual Review of Biomedical Engineering, 4,* 261–286.

Bennett, M. R., & Hasty, J. (2009). Microfluidic devices for measuring gene network dynamics in single cells. *Nature Reviews. Genetics, 10,* 628–638.

Bergmiller, T., Pena-Miller, R., Boehm, A., & Ackermann, M. (2011). Single-cell timelapse analysis of depletion of the universally conserved essential protein YgjD. *BMC Microbiology, 11*, 118.

Boedicker, J. Q., Li, L., Kline, T. R., & Ismagilov, R. F. (2008). Detecting bacteria and determining their susceptibility to antibiotics by stochastic confinement in nanoliter droplets using plug-based microfluidics. *Lab on a Chip, 8*(8), 1265–1272.

Boedicker, J. Q., Vincent, M. E., & Ismagilov, R. F. (2009). Microfluidic confinement of single cells of bacteria in small volumes initiates high-density behavior of Quorum sensing and growth and reveals its variability. *Angewandte Chemie (International Ed. in English), 48*(32), 5908–5911.

Burmølle, M., Ren, D., Bjarnsholt, T., & Sørensen, S. J. (2014). Interactions in multispecies biofilms: Do they actually matter? *Trends in Microbiology, 22*(2), 84–91.

Cabeen, M. T., & Jacobs-Wagner, C. (2005). Bacterial cell shape. *Nature Reviews. Microbiology, 3*, 601–610.

Campos, M., Surovtsev, I. V., Kato, S., Paintdakhi, A., Beltran, B., Ebmeier, S. E., et al. (2014). A constant size extension drives bacterial cell size homeostasis. *Cell, 159*, 1433–1446.

Caspi, Y. (2014). Deformation of filamentous Escherichia coli cells in a microfluidic device: A new technique to study cell mechanics. *PLoS One:*8e83775.

Cattoni, D. I., Fiche, J. B., Valeri, A., Mignot, T., & Nöllmann, M. (2013). Super-resolution imaging of bacteria in a microfluidics device. *PLoS One:*8e76268.

Chalfoun, J., Cardone, A., Dima, A. A., Allen, D. P., & Halter, M. W. (2010). Overlap-based cell tracker. *Journal of Research of the National Institute of Standards and Technology, 115*(6), 477–486.

Chen, Y. C., Allen, S. G., Ingram, P. N., Buckanovich, R., Merajver, S. D., & Yoon, E. (2015). Single-cell migration chip for chemotaxis-based microfluidic selection of heterogeneous cell populations. *Scientific Reports, 5*, 9980.

Cooper, S., & Helmstetter, C. E. (1968). Chromosome replication and the division cycle of Escherichia coli B/r. *Journal of Molecular Biology, 31*, 519–540.

Daniel, R. A., & Errington, J. (2003). Control of cell morphogenesis in bacteria: Two distinct ways to make a rod-shaped cell. *Cell, 113*, 767–776.

Danino, T., Mondragón-Palomino, O., Tsimring, L., & Hasty, J. (2010). A synchronized quorum of genetic clocks. *Nature, 463*, 326–330.

de Jong, I. G., Beilharz, K., Kuipers, O. P., & Veening, J. W. (2011). Live cell imaging of Bacillus subtilis and Streptococcus pneumoniae using automated time-lapse microscopy. *Journal of Visualized Experiments, 53*, 3145.

Donachie, W. D. (1968). Relationship between cell size and time of initiation of DNA replication. *Nature, 219*, 1077–1079.

Drescher, K., Dunkel, J., Cisneros, L. H., Ganguly, S., & Goldstein, R. E. (2011). Fluid dynamics and noise in bacterial cell-cell and cell-surface scattering. *Proceedings of the National Academy of Sciences of the United States of America, 108*(27), 10940–10945.

Ducret, A., Maisonneuve, E., Notareschi, P., Grosssi, A., Mignot, T., & Dukan, S. (2009). A microscope automated fluidic system to study bacterial processes in real time. *PLoS One:*4(9). e7282.

Ducret, A., Théodoly, O., & Mignot, T. (2013). Single cell microfluidic studies of bacterial motility. *Methods in Molecular Biology, 966*, 97–107.

Elgeti, J., Winkler, R. G., & Gompper, G. (2015). Physics of microswimmers—Single particle motion and collective behaviour: A review. *Reports on Progress in Physics, 78*(5), 056601.

Errington, J. (2015). Bacterial morphogenesis and the enigmatic MreB helix. *Nature Reviews Microbiology, 13*, 241–248.

Eun, Y. J., Utada, A. S., Copeland, M. F., Takeuchi, S., & Weibel, D. B. (2011). Encapsulating bacteria in agarose microparticles using microfluidics for high-throughput cell analysis and isolation. *ACS Chemical Biology, 6*(3), 260–266.

Gómez-Sjöberg, R., Morisette, D. T., & Bashir, R. (2005). Impedance microbiology-on-a-chip: Microfluidic bioprocessor for rapid detection of bacterial metabolism. *Journal of Microelectromechanical Systems, 14*, 829–838.

Grover, W. H., Ivester, R. H. C., Jensen, E. C., & Mathies, R. A. (2006). Development and multiplexed control of latching pneumatic valves using microfluidic logical structures. *Lab on a Chip, 6*, 623–631.

Guo, M. T., Rotem, A., Heyman, J. A., & Weitz, D. A. (2012). Droplet microfluidics for high-throughput biological assays. *Lab on a Chip, 12*, 2146–2155.

Hol, F. J. H., & Dekker, C. (2014). Zooming in to see the bigger picture: Microfluidic and nanofabrication tools to study bacteria. *Science, 346*, 1251821.

Huang, H. (2015). *Fluigi: An end-to-end software workflow for microfluidic design.* Boston University. PhD thesis.

Huang, S., Srimani, J. K., Lee, A. J., Zhang, Y., Lopatkin, A. J., Leong, K. W., et al. (2015). Dynamic control and quantification of bacterial population dynamics in droplets. *Biomaterials, 61*, 239–245.

Jeong, H.-H., Jin, S. H., Lee, B. J., Kim, T., & Lee, C.-S. (2015). Microfluidic static droplet array for analyzing microbial communication on a population gradient. *Lab on a Chip, 15*(3), 889–899.

Joyce, G., Robertson, B. D., & Williams, K. J. (2011). A modified agar pad method for mycobacterial live-cell imaging. *BMC Research Notes, 4*, 73.

Kaehr, B., & Shear, J. B. (2009). High-throughput design of microfluidics based on directed bacterial motility. *Lab on a Chip, 9*(18), 2632–2637.

Kim, H. J., Boedicker, J. Q., Choi, J. W., & Ismagilov, R. F. (2008). Defined spatial structure stabilizes a synthetic multispecies bacterial community. *Proceedings of the National Academy of Sciences of the United States of America, 105*(47), 18188–18193.

Kim, J., Hegde, M., & Jayaraman, A. (2010). Co-culture of epithelial cells and bacteria for investigating host–pathogen interactions. *Lab on a Chip, 10*, 43–50.

Kolderman, E., Bettampadi, D., Samarian, D., Dowd, S. E., Foxman, B., Jakubovics, N. S., et al. (2015). L-Arginine destabilizes oral multi-species biofilm communities developed in human saliva. *PLoS One, 10*(5). e0121835.

Lam, R. H. W., Kim, M. C., & Thorsen, T. (2009). Culturing aerobic and anaerobic bacteria and mammalian cells with a microfluidic differential oxygenator. *Analytical Chemistry, 81*(14), 5918–5924.

Larson, J. W., Yantz, G. R., Zhong, Q., Charnas, R., D'Antoni, C. M., Gallo, M. V., et al. (2006). Single DNA molecule stretching in sudden mixed shear and elongational microflows. *Lab on a Chip, 6*(9), 1187–1199.

Li, B., Qiu, Y., GLidle, A., McIlvenna, D., Luo, Q., Cooper, J., et al. (2014). Gradient microfluidics enables rapid bacterial growth inhibition testing. *Analytical Chemistry, 86*(6), 3131–3137.

Li, X., Yu, Z. T., Geraldo, D., Weng, S., Alve, N., Dun, W., et al. (2015). Desktop aligner for fabrication of multilayer microfluidic devices. *The Review of Scientific Instruments, 86*(7), 075008.

Lindner, A. B., Madden, R., Demarez, A., Stewart, E. J., & Taddei, F. (2008). Asymmetric segregation of protein aggregates is associated with cellular aging and rejuvenation. *Proceedings of the National Academy of Sciences of the United States of America, 105*, 3076–3081.

Liu, Y., & Singh, A. K. (2013). Microfluidic platforms for single-cell protein analysis. *Journal of Laboratory Automation, 18*, 446–454.

Longo, D., & Hasty, J. (2006). Dynamics of single-cell gene expression. *Molecular Systems Biology, 2*, 64.

Maisonneuve, E., & Gerdes, K. (2014). Molecular mechanisms underlying bacterial persisters. *Cell, 157*(3), 539–548.

Mannik, J., Driessen, R., Galajda, P., Keymer, J. E., & Dekker, C. (2009). Bacterial growth and motility in sub-micron constrictions. *Proceedings of the National Academy of Sciences of the United States of America, 106*, 14861–14866.

Mark, D., Haeberle, S., Roth, G., von Stetten, F., & Zengerle, R. (2010). Microfluidic lab-on-a-chip platforms: Requirements, characteristics and applications. *Chemical Society Reviews, 39*, 1153–1182.

Marsh, P. D. (1994). Microbial ecology of dental plaque and its significance in health and disease. *Advances in Dental Research, 8*, 263–271.

Marsh, P. D. (2005). Dental plaque: Biological significance of a biofilm and community lifestyle. *Journal of Clinical Periodontology, 32*(6), 7–15.

Mata, A., Fleischman, A. J., & Roy, S. C. (2005). Characterisation of polydimethylsiloxane (PDMS) properties for biomedical micro/nanosystems. *Biomedical Microdevices, 7*(4), 281–293.

Mathis, R., & Ackermann, M. (2016). Response of single bacterial cells to stress gives rise to complex history dependence at the population level. *Proceedings of the National Academy of Sciences of the United States of America, 113*, 4224–4229.

Matos, T., Senkbeil, S., Mendonça, A., Queiroz, J. A., Kutter, J. P., & Bulow, L. (2013). Nucleic acid and protein extraction from electropermeabilized E. coli cells on a microfluidics chip. *Analyst, 138*(24), 7347–7353.

McDonald, J. C., Duffy, D. C., Anderson, J. R., Chiu, D. T., Wu, H., Schueller, O. J., et al. (2000). Fabrication of microfluidic systems in poly(dimethylsiloxane). *Electrophoresis, 21*(1), 27–40.

Meijering, E., Dzyubachyk, O., & Smal, I. (2012). Methods for cell and particle tracking. *Methods in Enzymology, 504*, 183–200.

Miralles, V., Huerre, A., Malloggi, F., & Jullien, M. C. (2013). A review of heating and temperature control in microfluidic systems: Techniques and applications. *Diagnostics, 3*, 33–67.

Moffitt, J. R., Lee, J. B., & Cluzel, P. (2012). The single-cell chemostat: An agarose-based, microfluidic device for high-throughput, single-cell studies of bacteria and bacterial communities. *Lab on a Chip, 12*(8), 1487–1494.

Moolman, C. M., Huang, Z., Krishnan, S. T., Kerssemakers, J. W. J., & Dekker, N. H. (2013). Electron beam fabrication of a microfluidic device for studying submicron-scale bacteria. *Journal of Nanobiotechnology, 11*, 12.

Nance, W. C., Dowd, S. E., Samarian, D., Chludzinski, J., Delli, J., Battista, J., et al. (2013). A high-throughput microfluidic dental plaque biofilm system to visualize and quantify the effect of antimicrobials. *Journal of Antimicrobial Chemotherapy, 68*(11), 2550–2560.

Norman, T. M., Lord, N. D., Paulsson, J., & Losick, R. (2013). Memory and modularity in cell-fate decision making. *Nature, 503*, 481–486.

Ochs, C. J., Kasuya, J., Pavesi, A., & Kamm, R. D. (2014). Oxygen levels in thermoplastic microfluidic devices during cell culture. *Lab on a Chip, 14*, 459.

Oliveira, N. M., Martinez-Garcia, E., Xavier, J., Durham, W. M., Kolter, R., Kim, W., et al. (2015). Biofilm formation as a response to ecological competition. *PLoS Biology, 13*(7). e1002191.

Paintdakhi, A., Parry, B., Campos, M., Irnov, I., Elf, J., Surovtsev, I., et al. (2016). Oufti: An integrated software package for high-accuracy, high-throughput quantitative microscopy analysis. *Molecular Microbiology, 99*(4), 767–777.

Pamp, S. J., Harrington, E. D., Quake, S. R., Relman, D. A., & Blainey, P. C. (2012). Single-cell sequencing provides clues about the host interactions of segmented filamentous bacteria (SFB). *Genome Research, 22*(6), 1107–1119.

Pelletier, J., Halvorsen, K., Ha, B.-Y., Paparcone, R., Sandler, S., Woldringh, C., et al. (2012). Physical manipulation of the *Escherichia coli* chromosome reveals its soft nature. *Proceedings of the National Academy of Sciences of the United States of America, 109*, E2649–E2656.

Piccinini, F., Kiss, A., & Horvath, P. (2016). Cell tracker (not only) for dummies. *Bioinformatics, 32*(6), 955–957.

Piruska, A., Nikcevic, I., Lee, S. H., Ahn, C., Heineman, W. R., Limbach, P. A., et al. (2005). The autofluorescence of plastic materials and chips measured under laser irradiation. *Lab on a Chip, 5*(12), 1348–1354.

Qin, D., Xia, Y., & Whitesides, G. M. (2010). Soft lithography for micro- and nanoscale patterning. *Nature Protocols, 5*, 491–502.

Ramalho, T., Meyer, A., Mückl, A., Kapsner, K., Gerland, U., & Simmel, F. C. (2016). Single cell analysis of a bacterial sender-receiver system. *PLoS One, 11*(1) e0145829.

Regehr, K. J., Domenech, M., Koepsel, J. T., Carver, K. C., Ellison-Zelski, S. J., Murphy, W. L., et al. (2009). Biological implications of polydimethylsiloxane-based microfluidic cell culture. *Lab on a Chip, 9*, 2132–2139.

Rusconi, R., Garren, M., & Stocker, R. (2014). Microfluidics expanding the frontiers of microbial ecology. *Annual Review of Biophysics, 43*, 65–91.

Rusconi, R., Guasto, J. S., & Stocker, R. (2014). Bacterial transport suppressed by fluid shear. *Nature Physics, 10*(3), 212–217.

Saleh-Lakha, S., & Trevors, J. T. (2010). Perspective: Microfluidic applications in microbiology. *Journal of Microbiological Methods, 82*, 108–111.

Sauls, J. T., Li, D., & Jun, S. (2016). Adder and a coarse-grained approach to cell size homeostasis in bacteria. *Current Opinion in Cell Biology, 38*, 38–44.

Sharp, K. V., Adrian, R. J., Santiago, J. G., & Molho, J. I. (2005). Liquid flow in microchannels. In M. Gad-el-Hak (Ed.), *The MEMS handbook* (2nd ed., pp. 10-1–10-45). Boca Raton, FL: CRC Press.

Shih, S. C., Goyal, G., Kim, P. W., Koutsoubelis, N., Keasling, J. D., Adams, P. D., et al. (2015). A versatile microfluidic device for automating synthetic biology. *ACS Synthetic Biology, 16*(4), 1151–1164.

Si, F., Li, B., Margolin, W., & Sun, S. X. (2015). Bacterial growth and form under mechanical compression. *Science Reports, 5*, 11367.

Sikanen, T., Heikkilä, L., Tuomikoski, S., Ketola, R. A., Kostiainen, R., Franssila, S., et al. (2007). Performance of SU-8 microchips as separation devices and comparison with glass microchips. *Analytical Chemistry, 79*(16), 6255–6263.

Silhavy, T. J., Kahne, D., & Walker, S. (2010). The bacterial cell envelope. *Cold Spring Harbor Perspectives in Biology, 2*. a000414.

Sliusarenko, O., Heinritz, J., Emonet, T., & Jacobs-Wagner, C. (2011). High-throughput, subpixel precision analysis of bacterial morphogenesis and intracellular spatio-temporal dynamics. *Molecular Microbiology, 80*(3), 612–627.

Son, K., Brumley, D. R., & Stocker, R. (2015). Live from under the lens: Exploring microbial motility with dynamic imaging and microfluidics. *Nature Reviews. Microbiology, 13*, 761–775.

Song, Z., Borgwardt, L., Høiby, N., Wu, H., Sørenson, T. S., & Borgwardt, A. (2013). Prosthesis infection after orthopaedic joint replacement: The possible role of bacterial biofilms. *Orthopedic Review, 5*(2). e14.

Steinhaus, B., Garcia, M. L., Shen, A. Q., & Angenent, L. T. (2007). A portable anaerobic microbioreactor reveals optimum growth conditions for the methanogen *Methanosaeta concilii*. *Applied and Environmental Microbiology, 73*(5), 1653–1658.

Stewart, E. J., Madden, R., Paul, G., & Taddei, F. (2005). Aging and death in an organism that reproduces by morphologically symmetric division. *PLoS Biology, 3*. e45.

Streets, A. M., & Huang, Y. (2014). Microfluidics for biological measurements with single-molecule resolution. *Current Opinion in Biotechnology, 25*, 69–77.

Taheri-Araghi, S., Bradde, S., Sauls, J. T., Hill, N. S., Levin, P. A., Paulsson, J., et al. (2015). Cell-size control and homeostasis in bacteria. *Current Biology, 25*, 385–391.

Taheri-Araghi, S., Brown, S. D., Sauls, J. T., McIntosh, D. B., & Jun, S. (2015). Single-cell physiology. *Annual Review of Biophysics, 44*, 123–142.

Takeuchi, S., DiLuzio, W. R., Weibel, D. B., & Whitesides, G. M. (2005). Controlling the shape of filamentous cells of *Escherichia coli*. *Nano Letters, 5*, 1819–1823.

Taniguchi, Y., Choi, P. J., Li, G. W., Chen, H., Babu, M., Hearn, J., et al. (2010). Quantifying E. coli proteome and transcriptome with single-molecule sensitivity in single cells. *Science, 329*(5991), 533–538.

Tao, F. F., Xiao, X., Lei, K. F., & Lee, I. C. (2015). Paper-based cell culture microfluidics system. *BioChip Journal, 9*(2), 97–104.

Typas, A., Banzhaf, M., Gross, C. A., & Vollmer, W. (2012). From the regulation of peptidoglycan synthesis to bacterial growth and morphology. *Nature Reviews. Microbiology, 10*, 123–136.

Ullman, G., Wallden, M., Marklund, E. G., Mahmutovic, A., Razinkov, I., & Elf, J. (2013). High-throughput gene expression analysis at the level of single proteins using a microfluidic turbidostat and automated cell tracking. *Philosophical Transactions of the Royal Society of London. Series B, Biological Sciences, 368*. 20120025.

Unger, M. A., Chou, H.-P., Thorsen, T., Scherer, A., & Quake, S. R. (2000). Monolithic microfabricated valves and pumps by multilayer soft lithography. *Science, 288*, 113–116.

Velve-Casquillas, G., Le Berre, M., Piel, M., & Tran, P. T. (2010). Microfluidic tools for cell biological research. *Nano Today, 5*, 28–47.

Volpatti, L. R., & Yetisen, A. K. (2014). Commercialization of microfluidic devices. *Trends in Biotechnology, 32*, 347–350.

Voorn, W. J., & Koppes, L. J. (1998). Skew or third moment of bacterial generation times. *Archives of Microbiology, 169*, 43–51.

Voorn, W. J., Koppes, L. J., & Grover, N. B. (1993). Mathematics of cell division in Escherichia coli: Comparison between sloppy-size and incremental-size kinetics. *Current Topics in Molecular Genetics, 1*, 187–194.

Vulto, P., Dame, G., Maier, U., Makohilso, S., Podszun, S., Zahn, P., et al. (2010). A microfluidic approach for high efficiency extraction of low molecular weight RNA. *Lab on a Chip, 10*(5), 610–616.

Wakamoto, Y., Dhar, N., Chait, R., Schneider, K., Signorino-Gelo, F., Leibler, S., et al. (2013). Dynamic persistence of antibiotic-stressed mycobacteria. *Science, 339*(6115), 91–95.

Wallden, M., & Elf, J. (2011). Studying transcriptional interactions in single cells at sufficient resolution. *Current Opinion in Biotechnology, 22*, 81–86.

Wang, X., Kang, Y., Luo, C., Zhao, T., Liu, L., Jiang, X., et al. (2014). Heteroresistance at the single-cell level: Adapting to antibiotic stress through a population-based strategy and growth-controlled interphenotypic coordination. *mBio:5*(1)e00942-13.

Wang, P., Robert, L., Pelletier, J., Dang, W. L., Taddei, F., Wright, A., et al. (2010). Robust growth of Escherichia coli. *Current Biology, 20*, 1099–1103.

Weibel, D. B., Siegel, A. C., Lee, A., George, A. H., & Whitesides, G. M. (2007). Pumping fluids in microfluidic systems using the elastic deformation of poly(dimethylsiloxane). *Lab on a Chip, 7*, 1832–1836.

Wessel, A. K., Hmelo, L., Parsek, M. R., & Whiteley, M. (2013). Going local: Technologies for exploring bacterial microenvironments. *Nature Reviews. Microbiology, 11*, 337–348.

Westerwalbesloh, C., Grünberger, A., Stute, B., Weber, S., Wiechert, W., Kohlheyer, D., et al. (2015). Modeling and CFD simulation of nutrient distribution in picoliter bioreactors for bacterial growth studies on single-cell level. *Lab on a Chip, 15*, 4177–4186.

Whitesides, G. M. (2006). The origins and the future of microfluidics. *Nature, 442*, 368–373.

Whitesides, G. M., Ostuni, E., Takayama, S., Jiang, X., & Ingber, D. E. (2001). Soft lithography in biology and biochemistry. *Annual Review of Biomedical Engineering, 3*, 335–373.

Williams, M., Hoffman, M. D., Daniel, J. J., Madren, S. M., Dhroso, A., Korkin, D., et al. (2016). Short-stalked Prosthecomicrobium hirschii cells have a Caulobacter-like cell cycle. *Journal of Bacteriology, 198*, 1149–1159.

Wong, I., Atsumi, S., Huang, W. C., Wu, T. Y., Hanai, T., Lam, M. L., et al. (2010). An agar gel membrane-PDMS hybrid microfluidic device for long term single cell dynamic study. *Lab on a Chip, 10*, 2710–2719.

Wu, L. J., & Errington, J. (2011). Nucleoid occlusion and bacterial cell division. *Nature Reviews. Microbiology, 10*(1), 8–12.

Wu, F., van Schie, B. G., Keymer, J. E., & Dekker, C. (2015). Symmetry and scale orient Min protein patterns in shaped bacterial sculptures. *Nature Nanotechnology, 10*, 719–726.

Yehezkel, B., Rival, A., Raz, O., Cohen, R., Marx, Z., Camara, M., et al. (2016). Synthesis and cell-free cloning of DNA libraries using programmable microfluidics. *Nucleic Acids Research:44*(4)e35.

Young, K. D. (2006). The selective value of bacterial shape. *Microbiology and Molecular Biology Reviews, 70*, 660–703.

Young, K. D. (2007). Bacterial morphology: Why have different shapes? *Current Opinion in Microbiology, 10*, 596–600.

Young, J. W., Locke, J. C., Altinok, A., Rosenfeld, N., Bacarian, T., Swain, P. S., et al. (2012). Measuring single-cell gene expression dynamics in bacteria using fluorescence time-lapse microscopy. *Nature Protocols, 7*, 80–88.

Zhu, Y., & Fang, Q. (2013). Analytical detection techniques for droplet microfluidics—A review. *Analytica Chimica Acta, 17*(787), 24–35.

SECTION 2

Electron Microscopy

CHAPTER 4

Electron cryotomography

C.M. Oikonomou*, M.T. Swulius*, A. Briegel[†], M. Beeby[‡], Q. Yao*, G.J. Jensen*,[§],[1]

*California Institute of Technology, Pasadena, CA, United States
[†]Institute of Biology, Leiden University, Leiden, The Netherlands
[‡]Imperial College of London, London, United Kingdom
[§]Howard Hughes Medical Institute, Pasadena, CA, United States
[1]Corresponding author: e-mail address: jensen@caltech.edu

1 INTRODUCTION TO ELECTRON CRYOTOMOGRAPHY TECHNOLOGY

In biology, understanding is frequently limited by our ability to see. This problem is particularly acute for the smallest cells—Bacteria and Archaea. Light microscopy can resolve the general morphology of micron-scale cells, but not interior details on the scale of nanometres. Several techniques have been developed that can access these details, and they are the focus of this volume. In this chapter, we focus on electron cryotomography (ECT), a rapidly developing technique that allows us to image large macromolecular complexes directly in their native context inside intact cells. In essence, this technique, ECT, perhaps more than any other, provides us with the three-dimensional blueprints of bacterial cells.

Since its development in the 1930s, electron microscopy (EM) has delivered a wealth of information about subcellular structures. Not limited by the wavelength of light, EM can resolve details even to the atomic scale (Zemlin, Weiss, Schiske, Kunath, & Hermann, 1978). Practical considerations, however, limit its application to, and resolving power in, biological samples. For example, the column of the electron microscope is maintained at a high vacuum so if a sample of cells was inserted directly into the column, the water would instantly and violently bubble away with disastrous results. Traditional EM preparations circumvent this by chemically fixing, dehydrating and resin-embedding cells (Palade, 1952). While much information has been gained this way, many finer details are lost, and even some artefacts introduced (Pilhofer et al., 2014; Pilhofer, Ladinsky, McDowall, & Jensen, 2010).

In the 1980s, an alternative immobilization scheme was devised. Rather than chemically fixing and dehydrating cells, samples in standard aqueous media were simply frozen. Normal ice is a solid; in the freezing process, water molecules bond with their neighbours into a structured crystalline lattice. This lattice occupies a larger volume than the liquid, giving rise to the characteristic lower density of ice. On a macroscopic scale, this property allows life to survive in frozen-over lakes; on a microscopic scale, it destroys cellular structure. Very efficient cryogens, however, such as liquid ethane, can transfer heat out of a sample so rapidly that water molecules lose kinetic energy before they have a chance to find bonding partners. The resulting immobilized water is amorphous and not technically a solid and is therefore referred to as "vitreous" ice (Dubochet et al., 1988).

In cryogenic electron microscopy (cryo-EM), samples are rapidly plunged into, typically, liquid ethane to preserve native biological structures in vitreous ice. Samples are subsequently kept and imaged under cryogenic conditions (at or below the temperature of liquid nitrogen) to prevent the formation of crystalline ice. Plunge-frozen cells remain intact and are largely viable if thawed. Cryo-EM therefore captures a snapshot of a living cell at a single moment in time.

In the late 1990s, a specialized modality of cryo-EM was developed by incorporating tomography (Koster et al., 1997). In ECT, samples are imaged in a cryogenic transmission electron microscope (TEM). Many images of the same sample are collected, tilting the sample by a few degrees between each image. The result is a "tilt-series" of 2D projection images through the sample. Since each represents a different view of the same object, they can be digitally reconstructed into a 3D volume or "tomogram". The same principle is applied to medical imaging in CT scans (X-ray computed tomography). Fig. 1 shows a cryotomogram of a bacterial cell, in comparison to traditional thin-section EM.

One limitation of EM in its application to biological samples is the strong interaction of electrons with biological material. In phase-contrast TEM, image contrast arises from electrons that were elastically scattered by the sample interacting with the unscattered electron beam. If a sample is more than a few hundred nanometres thick, very few electrons emerge unscattered or singly scattered, yielding very little image information. Therefore, image quality is higher for thinner samples. For that reason, TEM samples are typically thinner than $\sim$500 nm. For such samples, the typical resolution is on the scale of large macromolecules, or a few nanometres. This sample thickness limit is the reason why ECT has been so widely and successfully applied to bacterial cell biology; many bacterial species are small enough to be imaged directly, in contrast to larger, e.g., eukaryotic, cells and organisms. In Section 3.1, we discuss methods that can be applied to render thicker samples accessible by ECT.

In summary, ECT produces 3D images of samples such as intact cells in an essentially native, "frozen-hydrated" state to $\sim$4 nm resolution—sufficient to see the shapes of large macromolecular complexes.

2 APPLICATIONS OF ECT TO MICROBIOLOGY

Though still in its infancy, ECT has already made several crucial contributions to bacterial cell biology. We will touch briefly on a few here, just to provide examples. First, ECT allowed the bacterial cytoskeleton to be visualized directly. Compared to

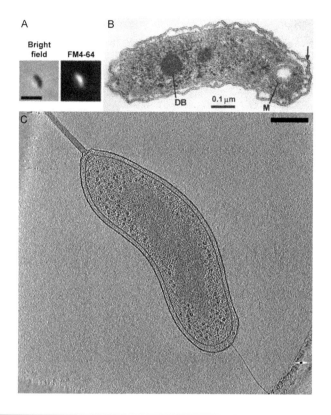

FIG. 1

Comparison of ECT with other techniques. *Bdellovibrio bacteriovorus* cells are shown, imaged by three different techniques: (A) phase-contrast and fluorescence light microscopy of a cell stained with FM4-64 membrane dye; (B) traditional thin-section TEM of fixed, dehydrated, resin-embedded and stained cells; and (C) ECT of native, frozen-hydrated cells. A central slice through the 3D tomogram is shown. Scale bars 2 μm (A), 100 nm (B) and 200 nm (C).

Panel (A): Reproduced from Fenton, A. K., Hobley, L., Butan, C., Subramaniam, S., & Sockett, R. E. (2010). A coiled-coil-repeat protein 'Ccrp' in Bdellovibrio bacteriovorus prevents cellular indentation, but is not essential for vibroid cell morphology. FEMS Microbiology Letters, 313(2), 89–95 with permission from FEMS (Oxford University Press). Panel (B): Reproduced from Burnham, J. C., Hashimoto, T., & Conti, S. F. (1968). Journal of Bacteriology, 96(4), 1366–1381 with permission from American Society for Microbiology. Panel (C): Reproduced from Oikonomou, C. M., & Jensen, G. J. (2016). A new view into prokaryotic cell biology from electron cryotomography. Nature Reviews Microbiology, 14(4), 205–220. doi: 10.1038/nrmicro.2016.7 with permission.

eukaryotic cells, bacteria were largely thought to lack cytoskeletal structures. Even though homologues of the three major classes of eukaryotic cytoskeletal elements (actin, tubulin and intermediate filaments) were found in bacterial genomes, very little evidence of filamentous structures was seen in decades of traditional TEM imaging. This turned out to be a negative side effect of the preparation methodologies. ECT imaging, with its improved preservation techniques, revealed the presence, and structure, of a wide array of cytoskeletal elements. Notable examples include MamK, an actin homologue that aligns microcompartments in magnetotactic bacteria (Komeili, Li, Newman, & Jensen, 2006; Scheffel et al., 2006); ParM, an actin homologue that segregates plasmids (Salje, Zuber, & Lowe, 2009); and FtsZ, a tubulin homologue that controls cell division in nearly all bacteria (Li, Trimble, Brun, & Jensen, 2007; Szwedziak, Wang, Bharat, Tsim, & Lowe, 2014). Bacteria even contain some filaments without known eukaryotic homologues, including Bactofilin, which controls cell shape (Kuhn et al., 2010). In an intriguing hint at how cytoskeletal filaments may have arisen in the first place, ECT helped show that a metabolic enzyme, CTP synthase, polymerizes into filaments that regulate cell shape in *Caulobacter crescentus* (Ingerson-Mahar, Briegel, Werner, Jensen, & Gitai, 2010). In fact, the diversity of filaments we have observed in cells by ECT has led us to expand the definition of the bacterial cytoskeleton to encompass rods, rings, twisted filament pairs, tubes, sheets, spirals, meshes and their combinations (Pilhofer & Jensen, 2013).

The power of ECT to image structures in their native context has helped reveal organizational details of bacterial cells, which were previously thought to be relatively undifferentiated. This organization includes highly ordered arrays of chemosensory components (Briegel et al., 2012; Liu et al., 2012) and even compartmentalization structures such as the bands that limit protein diffusion in *C. crescentus* stalks (Schlimpert et al., 2012). ECT also further characterized microcompartments for optimizing metabolism and storing nutrients (Beeby, Cho, Stubbe, & Jensen, 2012; Comolli, Kundmann, & Downing, 2006; Iancu et al., 2007, 2010; Konorty, Kahana, Linaroudis, Minsky, & Medalia, 2008; Psencik et al., 2009; Schmid et al., 2006; Ting, Hsieh, Sundararaman, Mannella, & Marko, 2007; Tocheva, Dekas, et al., 2013).

Finally, one of the greatest contributions of ECT to bacterial cell biology was to show us the astonishingly complex "nanomachines" that carry out highly specialized functions in cells (Alberts, 1998). These structures are typically composed of many copies of a dozen or more unique proteins and can span the cell envelope, translating cellular energy and signals in the cytoplasm into an effect in the extracellular environment or even another cell. Perhaps the best known of these nanomachines is the flagellar motor, which propels cells through liquid by rotating a long, thick filament (the flagellum). Since the motor is embedded in the cell envelope, it cannot be purified intact. ECT allowed us to see it at high resolution in situ and revealed many details of its protein architecture and adaptations in different species (Beeby et al., 2016; Chen et al., 2011; Liu et al., 2009; Murphy, Leadbetter, & Jensen, 2006). The structures of several other motility machines were also solved by ECT (Henderson & Jensen, 2006; Jasnin et al., 2013; Kurner, Frangakis, & Baumeister, 2005; Liu, McBride, & Subramaniam, 2007; Seybert, Herrmann, & Frangakis, 2006), including, notably, the type IVa pilus that *Myxococcus xanthus* cells use to glide along surfaces. In that case, analysis of deletion mutants allowed all of the components to be mapped, resulting in a complete architectural model (Chang et al., 2016) (Fig. 2).

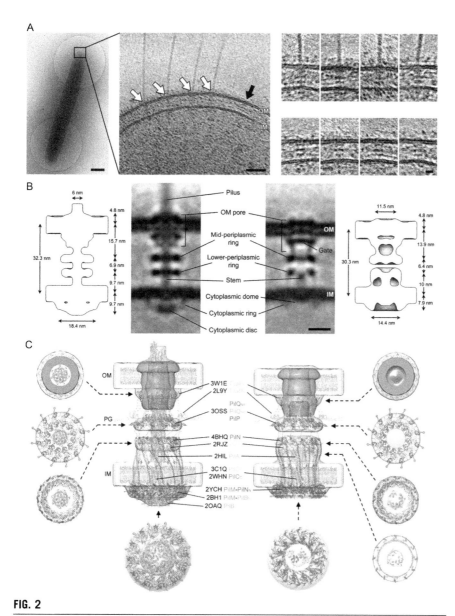

FIG. 2

ECT of a bacterial nanomachine. Here, ECT was used to dissect the architecture of the type IVa pilus machine in *Myxococcus xanthus*. (A) Cells were plunge-frozen and imaged by ECT to identify basal bodies either with (*white arrows* and top row at right) or without (*black arrow* and bottom row at right) pili assembled. *Asterisks* indicate characteristic basal body densities. The image at the left is a 2D projection image; the images at the right are slices from a tomogram. (B) Many copies of the structure (with and without an assembled pilus at left and right, respectively)

(Continued)

Other nanomachines exert their effects on other cells. For instance, ECT was used to solve the architecture of the type III secretion system, which many pathogenic bacteria use to inject toxins and other effector proteins into host cells (Abrusci et al., 2013; Hu et al., 2015). It also revealed the contractile mechanism by which the type VI secretion system launches toxin-tipped missiles into neighbouring competitors (Basler, Pilhofer, Henderson, Jensen, & Mekalanos, 2012).

3 COMPARISON TO OTHER TECHNIQUES

Experimental results must always be interpreted in the context of the technique used to produce them. Each technique is prone to its own set of characteristic errors, which must be controlled for. Moreover, for all the information a particular technique can provide, there is other information it cannot and which must be taken into account. Over the past 15 years, as we and others have applied ECT to a broad range of questions in microbiology, we have gained some insight into the strengths and limitations of ECT compared to other techniques. In many cases, results from different techniques have proved complementary. In a few cases, however, discrepancies are puzzling, requiring us to consider how the technical details of a method can influence its results. We therefore thought that this volume on imaging techniques offered an ideal opportunity to consider these technical limitations, as they relate to ECT. To that end, here we discuss several examples of studies that highlight certain disadvantages and advantages of ECT compared to other imaging methods.

3.1 LIMITATIONS OF ECT

3.1.1 Sample thickness

As mentioned earlier, a major limitation of ECT is the thickness of the sample. While some bacterial cells are narrow enough for high-resolution imaging, many are not, including common model organisms such as *Escherichia coli* and *Bacillus subtilis*. Other imaging modalities have no problem with cell thickness, and indeed, the relatively larger size of some bacterial species is an advantage for resolving structures by light microscopy. We have employed various strategies to get around this

---

FIG. 2—Cont'd   were used to generate subtomogram averages revealing higher-resolution details. (C) By repeating this process with a series of mutant strains lacking individual components, each protein's location was mapped in the structure, allowing available atomic models (shown with PDB identifiers) to be docked to create an architectural working model of the entire machine. Cross-sectional views of the complex at the indicated heights are shown at sides and bottom. *IM*, inner membrane; *OM*, outer membrane; *PG*, peptidoglycan layer. Scale bars 500 nm (A, left), 50 nm (A, middle), 5 nm (A, right) and 10 nm (B).

From Chang, Y. W., Rettberg, L., Treuner-Lange, A., Iwasa, J., Sogaard-Andersen, L., & Jensen, G. J. (2016). Architecture of the type IVa pilus machine. Science, 351 *(6278)*, aad2001. Reprinted with permission from AAAS.

thickness limit. For instance, in our studies of bacterial chemoreceptors, we used enzymes to introduce holes in thick cells, flattening them slightly for imaging (Briegel et al., 2009). In a similar approach, others used a phage lysis gene to flatten cell envelopes for ECT of chemoreceptor arrays (Fu et al., 2014). Since we were studying stable, membrane-bound structures, they remained intact, but this approach could not be used for other, e.g., cytoplasmic structures. Similarly, we and others have used minicells of various species to study chemoreceptor arrays (Briegel et al., 2012; Liu et al., 2012) and type III secretion systems (Hu et al., 2015). Produced by aberrant division near the cell poles, minicells are anucleate and nonreplicating and therefore cannot be used to study all cellular processes (Farley, Hu, Margolin, & Liu, 2016). For our studies of *B. subtilis* sporulation, we used a mutant lacking PonA (Tocheva, Lopez-Garrido, et al., 2013), which leads to thinner cells. However, as PonA functions in cell wall biogenesis, the use of a *ponA* mutant can be an important caveat to interpreting results.

There are also technological ways to circumvent cell thickness for ECT. One option is cryosectioning, in which a plunge-frozen sample is sliced with a diamond blade in a device called a cryo-ultramicrotome (Al-Amoudi, Norlen, & Dubochet, 2004). We used this approach to image cytoplasmic chemoreceptor arrays in *Rhodobacter sphaeroides* (Briegel et al., 2014). Cryosectioning can, however, introduce artefacts such as compression from the pressure of the blade (Al-Amoudi, Studer, & Dubochet, 2005; Dubochet et al., 2007). Recently, focused ion beam milling has been applied to ECT sample preparation. In this technique, a beam of gallium ions is used to ablate the top and bottom of a plunge-frozen cell sample, leaving a thin lamella that can be imaged at high resolution (Marko, Hsieh, Schalek, Frank, & Mannella, 2007). This technique introduces fewer artefacts and shows great promise in opening up cells of any size to interrogation by ECT.

3.1.2 Radiation damage

Another major limitation of ECT is radiation damage to the sample from imaging electrons breaking bonds, leading to charging and pressure that gradually destroy biological structures. In practice, this means that image quality degrades as more information is collected, so only a small electron dose can be used to interrogate biological samples. This is why ECT images have low contrast. This also puts a practical limit on the resolution that can be obtained for biological samples (on the order of 4 nm). While this resolution is high enough to reveal the shapes of large macromolecular complexes, it usually cannot show the finer details, such as the individual domains of the proteins they contain.

Our ECT imaging of MreB clearly illustrates this resolution limit. MreB, an actin homologue, was one of the first cytoskeletal proteins identified in bacteria and determines the rod shape of many species, including *B. subtilis* and *E. coli* (Jones, Carballido-Lopez, & Errington, 2001), through an as-yet-unclear mechanism. As we will discuss in more detail later, based on fluorescence microscopy, MreB was initially thought to form extended helices that wrapped around the cell, suggesting a global role in directing sites of cell wall synthesis to maintain rod shape. However, in our ECT imaging of 18 different rod-shaped bacterial species (~6000 tomograms) at that point, we had never observed a long, helical filament associated with the membrane. In fact, we observed no membrane-associated filaments of any conformation that seemed likely to be MreB, even using custom software to search for filaments in tomograms of cells from six different rod-shaped species (Swulius et al., 2011) (Fig. 3A–D). We knew that we could resolve filaments; Jan Löwe's group had used ECT to see MreB filaments in vitro, proving their resolvability (Salje, van den Ent, de Boer, & Lowe, 2011), and we had observed several other filaments with similar dimensions in our tomograms (including MreB bundles in the cytoplasm). So we used control experiments, introducing artificial filaments into our tomograms digitally, to determine the maximum length of an MreB-like filament that could go undetected in our analysis. The result told us that whatever structure MreB was forming in vivo, it had to be less than 80 nm in length, too short to be organizing cell wall synthesis along a cell-encompassing scaffold (Swulius et al., 2011). Thus, while our results ruled out some possibilities, the native structure of MreB and the mechanism by which it determines rod shape remain a puzzle, which we hope improvements in the resolution of ECT in the future will solve.

One tool to increase the resolution of ECT is subtomogram averaging. If a certain cellular structure has a fairly rigid, invariant structure and many copies of it can be identified (either in multiple copies in a cell or in different cells in multiple tomograms), these copies can be computationally averaged. Invariant regions of the structure reinforce one another, while variable background gets averaged out, increasing the signal-to-noise ratio and yielding higher, even subnanometre, resolution. Still, in many cases, the resolution is insufficient to see details of proteins. For instance, we used subtomogram averaging to produce a high-resolution map of the highly ordered arrays of bacterial chemosensory proteins inside cells, but information from X-ray crystallography was still needed to build pseudoatomic models (Briegel et al., 2012; Cassidy et al., 2015) (Fig. 4).

3.1.3 The missing wedge
ECT data is three dimensional, but it is not fully isotropic. This is due to the geometry of the technique. In order to obtain information from all angles, the sample would have to be tilted a full 180 degrees. As the sample is tilted to higher angles, however, it becomes effectively thicker (think of tilting a long and thin horizontal bar in a vertical beam). So instead of the full −90 degrees to +90 degrees tilt range, ECT is limited in practice to about −60 degrees to +60 degrees. This results in a "missing wedge" of information in the reconstruction, an anisotropy that decreases the resolution of tomograms in the z-direction (parallel to the imaging beam).

The effects of the missing wedge can be illustrated with two examples. One concerns the chemosensory arrays described earlier. In 2011, an ECT study of *E. coli* reported that chemoreceptors are packed more or less tightly in arrays depending on nutrient availability (Khursigara et al., 2011). This result was contradicted by subsequent ECT imaging of *C. crescentus* showing that strict hexagonal spacing of receptors in arrays is invariant in different growth conditions and activation states

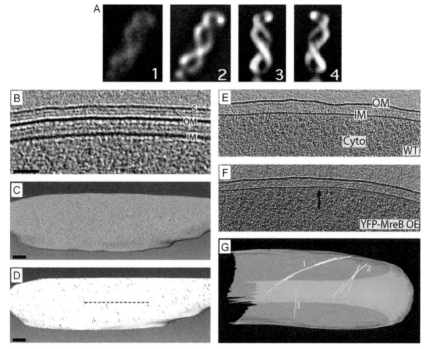

FIG. 3

ECT and the structure of MreB. (A) Based on light microscopy of fluorescently tagged versions, MreB was thought to form an extended helix wrapping around the cell, as seen in this *E. coli* cell expressing YFP-MreB. (1) and (2) show a single optical section before and after deconvolution, respectively, and (3) and (4) show rotated 3D reconstructions of the same cell using all sections. (B) However, no filaments were apparent close to the membrane in ECT images of wild-type *C. crescentus* cells containing unlabeled MreB, as shown in this tomographic slice. Computational search methods similarly failed to detect such filaments. (C) Projections of all density from the periplasmic face of the inner membrane extending 13 nm into the cytoplasm (higher density shown in *red*). (D) The results of a segment-based search for filaments up to 21 nm away from the inner membrane (potential filaments shown in *red*). The *dashed line* indicates the location of the tomographic slice shown in (B). (E, F) While no filaments were observed in wild-type *E. coli* cells either, they were apparent (*arrow*) in cells overexpressing the same YFP-MreB construct visualized in (A) using the same induction protocol. (G) A 3D segmentation of one such cell with helical filaments indicated. Scale bars 50 nm (B) and 100 nm (C,D).

Panel (A): Reproduced from Shih, Y. L., Le, T., & Rothfield, L. (2003). Division site selection in Escherichia coli involves dynamic redistribution of Min proteins within coiled structures that extend between the two cell poles. Proceedings of the National Academy of Sciences of the United States of America, 100(13), 7865–7870 with permission from National Academy of Sciences, U.S.A. Copyright (2003). Panels (B–D): Reproduced from Swulius, M. T., Chen, S., Jane Ding, H., Li, Z., Briegel, A., Pilhofer, M., ... Jensen, G. J. (2011). Long helical filaments are not seen encircling cells in electron cryotomograms of rod-shaped bacteria. Biochemical and Biophysical Research Communications, 407(4), 650–655. doi:10.1016/j.bbrc.2011.03.062 with permission from Elsevier. Panels (E–G): Reproduced from Swulius, M. T., & Jensen, G. J. (2012). The helical MreB cytoskeleton in Escherichia coli MC1000/pLE7 is an artifact of the N-terminal yellow fluorescent protein tag. Journal of Bacteriology, 194(23), 6382–6386. doi:10.1128/JB.00505-12 with permission from the American Society for Microbiology.

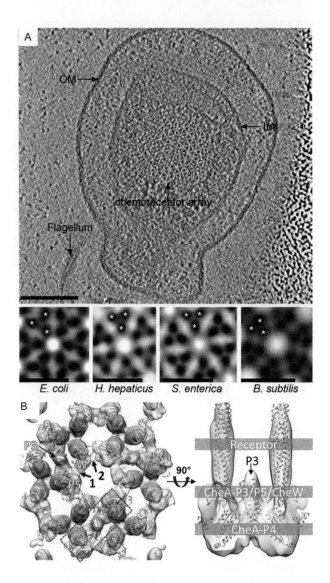

FIG. 4

Solving the in vivo architecture of the bacterial chemoreceptor array. (A) ECT was used to reveal the hexagonal lattice architecture of bacterial chemosensory arrays. At top is a tomographic slice through an *E. coli* minicell showing a top view of the array. At bottom are subtomogram averages from several species, showing the conserved hexagonal architecture of repeating trimeric subunits (*asterisks*; each corresponds to a chemoreceptor dimer). Scale bars 100 nm (*top*) and 12 nm (*bottom*). (B) This information was combined with X-ray crystallography to reveal the finer details of protein location within the array. Shown here is a pseudoatomic model mapping the location of key protein domains in the baseplate by docking crystal structures into a map produced by subtomogram averaging. P3 and P5 are two domains of the kinase CheA, and CheW is a coupling protein.

Panel (A): Reproduced from Briegel, A., Li, X., Bilwes, A. M., Hughes, K. T., Jensen, G. J., & Crane, B. R. (2012). Bacterial chemoreceptor arrays are hexagonally packed trimers of receptor dimers networked by rings of kinase and coupling proteins. Proceedings of the National Academy of Sciences of the United States of America, 109*(10), 3766–3771. doi:10.1073/pnas.1115719109 with permission. (B) Reproduced from Cassidy, C. K., Himes, B. A., Alvarez, F. J., Ma, J., Zhao, G., Perilla, J. R., . . . Zhang, P. (2015). CryoEM and computer simulations reveal a novel kinase conformational switch in bacterial chemotaxis signaling.* eLife, 4. *doi:10.7554/ eLife.08419, under CC BY 4.0 license.*

(Briegel, Beeby, Thanbichler, & Jensen, 2011). The earlier result was likely due to the missing wedge in imaging. The chemoreceptor array was visualized from the side and then rotated 90 degrees to examine the arrangement from a top view. This meant that distances between receptors were measured in the z-direction, where anisotropic smearing of densities can lead to error. The later ECT results that showed no difference in packing between nutrient conditions bypassed the limitation of the missing wedge by taking the measurements in nonrotated top views (i.e. in the x,y-plane) (Briegel et al., 2011).

The second example involves the bacterial division protein FtsZ. As we described earlier, ECT imaging identified native FtsZ filaments in bacterial cells, but the mechanism by which FtsZ drives cell constriction is still a matter of debate. One model proposes that individual filaments can pull the membrane through conformational changes (Lu, Reedy, & Erickson, 2000). Another model proposes that filaments slide against one another in an overlapping ring (Lan, Daniels, Dobrowsky, Wirtz, & Sun, 2009). The first ECT images of FtsZ in a small number of cells showed both isolated filaments and more complete bundles (Li et al., 2007). A later paper with many more and higher quality ECT images showed complete bundles (Szwedziak et al., 2014). While these studies were quite informative, the missing wedge complicated interpretation of the ECT results because filaments can only be followed about two-thirds of the way around the cell; the remaining one-third is blurred out as a result of the missing angular information (Fig. 5A–C).

3.1.4 Identification
A major challenge in EM studies is identifying a structure of interest. Unless it has a characteristic localization or a characteristic shape, it can be difficult or impossible to find within a visually crowded tomogram of a bacterial cell.

Immunolabelling with gold-conjugated antibodies can identify structures on the surface of an intact cell, but not inside. And while attempts are being made to develop genetically encoded EM-visible tags (Diestra, Fontana, Guichard, Marco, & Risco, 2009; Mercogliano & DeRosier, 2007; Wang, Mercogliano, & Lowe, 2011), there is as yet no practical option comparable to GFP for light microscopy. Instead, many proteins are located with the help of correlated light and electron microscopy (CLEM), in which fluorescent proteins are detected by light microscopy and the corresponding locations subsequently imaged at high resolution by ECT. This still leaves the problem of identifying a novel structure when it is first observed in a tomogram.

Our ECT imaging of the bacterial type VI secretion system perhaps best illustrates this problem. After we discovered the characteristic tubes in our tomograms of *Vibrio cholerae* cells in 2007, we first suspected they were cytoskeletal elements. Lacking any quick way to identify them, we did not pursue it beyond showing the pictures to other microbiologists. Eventually, this led us to John Mekalanos, who was working on the type VI secretion system, a conserved virulence factor his group identified in 2006 (Pukatzki et al., 2006). Together guessing that our tubes might be the product of the genes he was studying, we started working to test this idea, using

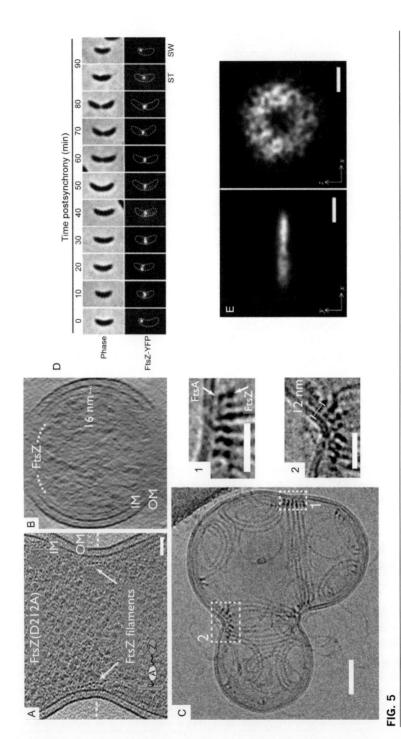

FIG. 5

See figure legend on opposite page

CLEM to confirm the identification and reveal the mechanism of the system (Basler et al., 2012) (Fig. 6). Interestingly, it appears that these structures may have been seen in multiple bacterial species by traditional EM since the 1960s, but nobody knew what they were (Bonemann, Pietrosiuk, & Mogk, 2010).

3.1.5 Static snapshots

Another limitation of ECT is that it records only frozen images, essentially snapshots. This can be problematic when interrogating dynamic processes. Cell constriction, for instance, is a highly dynamic process. Super-resolution light microscopy results suggest that the half-life of individual FtsZ monomers in filaments is only 150 ms (Biteen, Goley, Shapiro, & Moerner, 2012). ECT studies of FtsZ capture only relatively few temporal snapshots of this dynamic process, so we can build up only a patchy time-course of events, rather than watching the process unfold continuously in real time.

As another example, chemosensory arrays are inherently dynamic, changing their signalling state in response to chemical cues. In order to investigate whether the

FIG. 5

ECT and the structure of FtsZ filaments. (A) ECT was used to reveal the architecture of FtsZ inside intact cells, as shown here for a dividing *E. coli* cell. To increase the number of filaments, a stabilized variant of FtsZ (D212A) was overexpressed in these cells. Arrows in the tomographic slice indicate cross-sections of FtsZ filaments lining the division plane.
(B) A rotated slice from the same tomogram, showing the division plane in cross-section. FtsZ filaments can be seen, but the missing wedge of information at top and bottom prevents us from following individual filaments around the full circumference of the cell.
(C) Structures formed by purified FtsZ on liposomes in vitro. Note how easy it is to see the filaments here, compared to the cell cytosol in (A). Densities between FtsZ and the membrane corresponding to FtsA can be clearly seen, as illustrated by the enlargements in (1) and (2). While the relative protein levels of FtsA and FtsZ are likely different in this context, the location of FtsA in vivo can only be inferred, as shown by the cartoon in (A).
(D) The nonphysiological timing of FtsZ-YFP localization to the midplane when overexpressed as a fusion protein from a plasmid. FtsZ is normally cell cycle regulated. ST and SW indicate stalked and swarmer cells, respectively. (E) The ~250 nm thick ring of fluorescent signal from FtsZ-Dendra2 at the midplane of a *C. crescentus* cell. Shown are 2D projections from super-resolution 3D imaging along the *y*-axis (*left*) and in cross-sectional view (*right*). Scale bars 50 nm (A–C), 25 nm (1,2) and 200 nm (E).
Panels (A–C): Reproduced from Szwedziak, P., Wang, Q., Bharat, T. A., Tsim, M., & Lowe, J. (2014). Architecture of the ring formed by the tubulin homologue FtsZ in bacterial cell division. eLife, 4. doi:10.7554/eLife.04601 under CC BY 4.0 license. Panel (D): Reproduced from Goley, E. D., Yeh, Y. C., Hong, S. H., Fero, M. J., Abeliuk, E., McAdams, H. H., & Shapiro, L. (2011). Assembly of the Caulobacter cell division machine. Molecular Microbiology, 80(6), 1680–1698. doi:10.1111/j.1365-2958.2011.07677.x with permission. © 2011 Blackwell Publishing Ltd. (E) Reproduced from Biteen, J. S., Goley, E. D., Shapiro, L., & Moerner, W. E. (2012). Three-dimensional super-resolution imaging of the midplane protein FtsZ in live Caulobacter crescentus cells using astigmatism. Chemphyschem, 13(4), 1007–1012. doi:10.1002/cphc.201100686 with permission.
© 2012 Wiley-VCH Verlag GmbH & Co. KGaA, Weinheim.

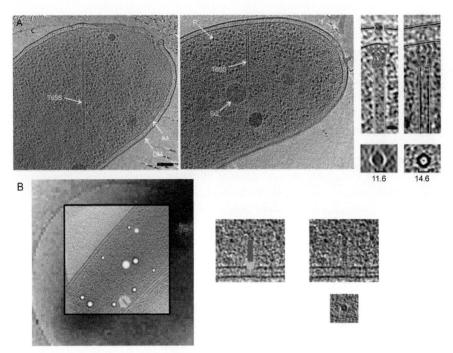

FIG. 6

Identifying the type VI secretion system. (A) The structure of the type VI secretion system in its extended (prefiring; *left*) and contracted (postfiring; *middle*) conformations as seen by ECT in intact *Vibrio cholerae* cells. Features of the two conformations are highlighted at right in side (*top*) and cross-sectional (*bottom*) views. (B) Super-resolution CLEM was used to identify type VI secretion system structures in *Myxococcus xanthus*. Cryogenic photoactivated localization microscopy was used to locate VipA-PA-GFP signal (*red/yellow*) and identified locations were then imaged by ECT (tomographic slice overlaid). This method identified a short assembly intermediate of the secretion system, shown segmented in *blue* and in longitudinal and cross-sectional views at right. Scale bars 100 nm (A, *left*) and 20 nm (A, *right*).

Panel (A): Reproduced from Basler, M., Pilhofer, M., Henderson, G. P., Jensen, G. J., & Mekalanos, J. J. (2012). *Type VI secretion requires a dynamic contractile phage tail-like structure.* Nature, 483*(7388)*, 182–186. doi:10.1038/nature10846 with permission. Panel (B): Reproduced from Chang, Y. W., Chen, S., Tocheva, E. I., Treuner-Lange, A., Lobach, S., Sogaard-Andersen, L., & Jensen, G. J. (2014). *Correlated cryogenic photoactivated localization microscopy and cryo-electron tomography.* Nature Methods, 11*(7)*, 737–739. doi:10.1038/nmeth.2961 with permission.

signalling state alters the structure of the arrays by ECT, we used either timed freezing after exposure to an attractant or a genetic approach, taking advantage of regulatory protein deletions that lock the receptors in different signalling states (Briegel et al., 2011, 2013).

3.2 ADVANTAGES OF ECT
3.2.1 Native proteins

One of the principal advantages of ECT compared to other imaging techniques is that it can image native proteins in fully wild-type cells. In contrast, fluorescence light microscopy relies on protein tags, which can disrupt the function and localization of the carrier (Landgraf, Okumus, Chien, Baker, & Paulsson, 2012). Indeed, in *C. crescentus* the localization of *most* proteins is disrupted by a terminal fluorescent tag (only 126 of 352 proteins showed the same localization when tagged on the N-terminus as on the C-terminus) (Werner et al., 2009). In addition, fluorescent fusion proteins are commonly expressed exogenously, altering expression levels, timing and, potentially, topography. Each of these factors can introduce artefacts, as we illustrate here with two (by now familiar) examples.

As discussed earlier, it used to be thought that MreB forms long helical cables just inside the cell membrane. This result was based on light microscopy of fluorescent fusions of MreB in *B. subtilis* (Jones et al., 2001), *E. coli* (Shih, Le, & Rothfield, 2003) and other rod-shaped cells. ECT imaging showed that MreB structures, if filamentous at all, are significantly shorter (<80 nm) and so do not encircle cells (Swulius et al., 2011). To investigate whether the fluorescent tag could be altering the structure of MreB, we imaged *E. coli* cells expressing the N-terminal YFP-MreB construct that had previously shown helical localization (Shih et al., 2003) and found that the tag did in fact induce helical filaments not seen in wild-type cells (Swulius & Jensen, 2012) (Fig. 3E–G). Like many other tagged protein variants, this fluorescent fusion of MreB is only partially functional (Shih et al., 2003). Why it forms extended helices remains unknown; it could be due to multimerization of the fluorescent protein (Landgraf et al., 2012), mislocalization arising from altered MreB function, or unnatural expression level. It was later shown that the N-terminus of MreB faces the membrane (Salje et al., 2011), so N-terminal fusions of this protein may be particularly disruptive of native localization.

An interesting caveat about data interpretation also arises from this study. While some fluorescent fusions of MreB, such as the N-terminal fusion in *E. coli*, form de facto helices, other fusion protein structures may not have been helical at all, but only interpreted that way. In some cases, puncta were observed along the edges of the cell, which may have been connected into apparent helices by the human eye or by excessive image deconvolution, as pointed out by Chastanet and Carballido-Lopez (2012). More recent light microscopy results suggest that fluorescent MreB structures are highly dynamic spots no larger than the diffraction limit (consistent with ECT results) that travel around the circumference of the cell (Dominguez-Escobar et al., 2011; Garner et al., 2011; van Teeffelen et al., 2011). Another possible explanation for earlier interpretation of helical structures in live-cell imaging may be that relatively long exposure times turned fast-moving discrete spots into continuous streaks.

Imaging fluorescent fusion proteins has also added to confusion about the mechanism of FtsZ activity. Like MreB, FtsZ-GFP is nonfunctional, requiring an additional untagged copy for cell viability (Ma, Ehrhardt, & Margolin, 1996). Exogenous expression can lead to expression level and timing artefacts, particularly for cell cycle-regulated proteins such as FtsZ. Highly expressed plasmid-encoded FtsZ-YFP produces a ring-like signal at the mid-cell early in the cell cycle, in fact well before the native protein even begins to be expressed (Goley et al., 2011; Fig. 5D). This and other results reinforced the idea that a complete ring of filaments mediates cell constriction, even in the early stages of cell division. ECT imaging, however, showed natively expressed wild-type FtsZ forms incomplete rings that can accompany constriction (Li et al., 2007), although at later stages of constriction, filaments overlap in most cells to form a ring-like bundle that extends all the way around the division plane in most cells (Szwedziak et al., 2014).

One caveat to keep in mind when interpreting fluorescence microscopy is that it is not possible to distinguish between a signal produced by concentrated but disconnected proteins and a signal produced by an actual filament. Given the extremely rapid turnover of individual FtsZ monomers in filaments (Biteen et al., 2012), it is likely that much of the band-like FtsZ signal seen in cells by light microscopy arises from dissociated proteins still localized to the midplane. In fact, super-resolution light microscopy found that the FtsZ "ring" extends radially by approximately 250 nm into the interior of the cell, and the authors reported this to be a reliable measurement, not just a blur resulting from limited resolution (Biteen et al., 2012; Fig. 5E). ECT imaging of native cells has never observed FtsZ *filaments* anywhere other than close ($\sim$15 nm) to the membrane (Li et al., 2007; Szwedziak et al., 2014). Therefore, the fluorescence signal likely reflects transiently dissociated proteins in a dynamic system.

3.2.2 Intact cells
Another strength of ECT is that it can be applied to *intact* cells in a *fully hydrated* state. Other methods such as cryo-EM single particle reconstruction or X-ray crystallography require macromolecular complexes to be purified out of the cell. In some cases, this alters the structures of the proteins. This problem is particularly acute for complexes containing hydrophobic regions, such as transmembrane proteins. We observed the effects of purification on one such membrane-spanning complex—the type IVa pilus nanomachine described earlier. While subcomplexes of the machine had been previously visualized by cryo-EM single particle reconstruction, we found that the detergent solubilization needed for purification significantly altered the structures compared with the in vivo conformations seen by ECT (Chang et al., 2016; Fig. 7).

Similarly, for bacterial chemosensory arrays, early models proposed for the arrangement of their components in the cell were based on the structures of crystallized components (Park et al., 2006; Shimizu et al., 2000). The full arrays, embedded in the cell envelope, could not be purified intact. Thus it was not until the architecture could be seen in situ by ECT that the true arrangement was discovered (Briegel et al., 2012; Liu et al., 2012).

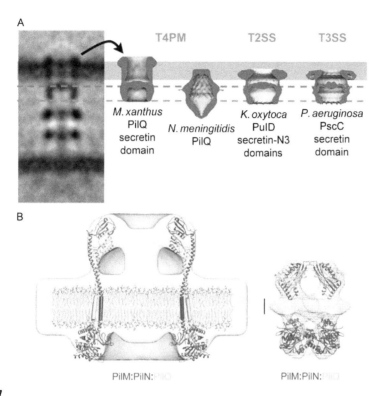

FIG. 7

Differences in structures solved in vivo and in vitro. (A) A comparison of the structure of the secretin domain of PilQ solved in vivo by subtomogram averaging of a ΔtsaP mutant of *M. xanthus* (*left*) and homologous secretin domains solved in vitro by cryo-EM single particle reconstruction of detergent-solubilized purified proteins (*right*). T4PM, type IV pilus machine; T2SS, type II secretion system; T3SS, type III secretion system. *Gray bar* indicates outer membrane. (B) A comparison of the structure of the type IV pilus subcomplex comprising PilM, PilN and PilO solved in vivo by subtomogram averaging (*left*) and in vitro by cryo-EM single particle reconstruction after detergent solubilization (*right*).

From Chang, Y. W., Rettberg, L., Treuner-Lange, A., Iwasa, J., Sogaard-Andersen, L., & Jensen, G. J. (2016). Architecture of the type IVa pilus machine. Science, 351(6278), aad2001. Reprinted with permission from AAAS.

3.2.3 3D

The final advantage of ECT that we will discuss here is its power to image in three dimensions. As we discussed earlier, this was invaluable in determining that MreB does not form extended helices around the cell, and it also allowed us to create architectural models of bacterial nanomachines. Another example that illustrates this advantage involves the organization of the multigigadalton peptidoglycan sacculus that forms the cell wall of most bacteria (Sobhanifar, King, & Strynadka, 2013).

The sacculus is made up of long rigid glycan polymers cross-linked by short peptides. The arrangement of these basic building blocks, however, has long been a subject of debate. The heavily favoured model proposed that glycan strands run circumferentially around a rod-shaped bacterial cell (Holtje, 1998), but others argued that they stand perpendicular to the surface, cross-linked into a lattice (Dmitriev et al., 2003). In sacculi purified from Gram-negative bacteria, ECT showed that the main pattern of density was long, thin elements running side by side around the cell, confirming the circumferential model (Gan, Chen, & Jensen, 2008).

Atomic force microscopy (AFM) of dehydrated *B. subtilis* sacculi suggested a vastly different peptidoglycan arrangement in Gram-positive cells. AFM images showed periodic structures that were interpreted to be the surface of parallel cables ~50 nm wide formed by tightly wound peptidoglycan coils (Hayhurst, Kailas, Hobbs, & Foster, 2008). The model was that glycan strands first wound tightly into 50 nm coils, and then these coils wrapped around the circumference of the cell. AFM, however, can only record the topology of a sample, not what lies beneath. So to obtain a true 3D view of the system, we turned again to ECT. Unfortunately, we found that sacculi purified from Gram-positive bacteria such as *B. subtilis* were thicker than those from Gram-negative cells and so ECT could not resolve individual strands directly. Still, ECT imaging gave us decisive clues about the native structure of the sacculus, notably that it exhibited continuous density when viewed from the side. This result is inconsistent with parallel cylindrical cables. We did observe furrows on the surface of the sacculus, as seen by AFM, but they were widely spaced and appeared to be the result of the degradation of layers of peptidoglycan during cell growth. Overall, our ECT imaging showed that the basic architecture of the

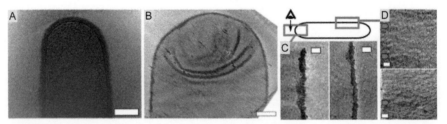

FIG. 8

ECT of the Gram-positive cell wall. ECT was used to investigate the structure of peptidoglycan in intact Δ*ponA* mutant *B. subtilis* cells (A) and sacculi purified from wild-type *B. subtilis* (B). As seen in top–down views of tomographic slices perpendicular (C) and parallel (D) to the long axis of the sacculus, the peptidoglycan is uniformly dense, incompatible with coiled cables. *Red arrows* in (D) indicate surface texture likely caused by degradation of the outermost peptidoglycan layer during growth, corresponding to the furrows seen in AFM images and interpreted to be the surface of cables. Scale bars 200 nm (A), 250 nm (B) and 50 nm (C,D).

Panels (A–D): Reproduced from Beeby, M., Gumbart, J. C., Roux, B., & Jensen, G. J. (2013). Architecture and assembly of the Gram-positive cell wall. Molecular Microbiology, 88(4), 664–672. doi:10.1111/mmi.12203 with permission. © 2013 John Wiley & Sons Ltd.

Gram-positive cell wall is in fact no different than that of the Gram-negative cell wall; it simply has additional layers of circumferential peptidoglycan (Beeby, Gumbart, Roux, & Jensen, 2013; Fig. 8). This conclusion was bolstered by ECT imaging of sporulation in *B. subtilis*, which showed that the peptidoglycan thins to a single layer during the process of engulfment, which is incompatible with any other model (Tocheva, Lopez-Garrido, et al., 2013).

CONCLUSIONS AND FUTURE DIRECTIONS

The past 15 years of ECT imaging have given us an unparalleled window onto native bacterial cell structure. We have seen the diversity of the bacterial cytoskeleton, observed the organization of intact cells and looked at the complex nanomachines that carry out jobs ranging from motility to infection. But even with these pieces in hand, we still have a considerable way to go before we have a complete understanding of the architecture of the bacterial cell. For instance, as we discussed earlier, we still do not know the constrictive mechanism of FtsZ in cell division, or the structure and mechanism of MreB in cell shape determination. Ultimately, we want to visualize all the large complexes in a cell and understand their interactions. Ongoing technology development continues to improve the resolution of ECT, and the utility of CLEM to direct it. Better segmentation and analysis tools are also being developed to analyze tomograms for the arrangement of structures in intact cells. Together, these improvements will help bring us closer to a complete understanding of bacterial cells.

ACKNOWLEDGEMENTS

This work was supported by Biotechnology and Biological Sciences Research Council grant BB/L023091/1 to M.B. Microbiological research in the Jensen laboratory is supported by the NIH (R01 GM101425 to G.J.J.), the Howard Hughes Medical Institute, the Beckman Institute at Caltech, Caltech's Center for Environmental Microbial Interactions, gifts to Caltech from the Gordon and Betty Moore Foundation and the Agouron Institute, and the John Templeton Foundation as part of the Boundaries of Life project. The opinions expressed in this publication are those of the authors and do not necessarily reflect the views of the John Templeton Foundation.

REFERENCES

Abrusci, P., Vergara-Irigaray, M., Johnson, S., Beeby, M. D., Hendrixson, D. R., Roversi, P., ... Lea, S. M. (2013). Architecture of the major component of the type III secretion system export apparatus. *Nature Structural & Molecular Biology, 20*(1), 99–104. http://dx.doi.org/10.1038/nsmb.2452.

Al-Amoudi, A., Norlen, L. P., & Dubochet, J. (2004). Cryo-electron microscopy of vitreous sections of native biological cells and tissues. *Journal of Structural Biology, 148*(1), 131–135. http://dx.doi.org/10.1016/j.jsb.2004.03.010.

Al-Amoudi, A., Studer, D., & Dubochet, J. (2005). Cutting artefacts and cutting process in vitreous sections for cryo-electron microscopy. *Journal of Structural Biology, 150*(1), 109–121. http://dx.doi.org/10.1016/j.jsb.2005.01.003.

Alberts, B. (1998). The cell as a collection of protein machines: Preparing the next generation of molecular biologists. *Cell, 92*(3), 291–294.

Basler, M., Pilhofer, M., Henderson, G. P., Jensen, G. J., & Mekalanos, J. J. (2012). Type VI secretion requires a dynamic contractile phage tail-like structure. *Nature, 483*(7388), 182–186. http://dx.doi.org/10.1038/nature10846.

Beeby, M., Cho, M., Stubbe, J., & Jensen, G. J. (2012). Growth and localization of polyhydroxybutyrate granules in Ralstonia eutropha. *Journal of Bacteriology, 194*(5), 1092–1099. http://dx.doi.org/10.1128/JB.06125-11.

Beeby, M., Gumbart, J. C., Roux, B., & Jensen, G. J. (2013). Architecture and assembly of the Gram-positive cell wall. *Molecular Microbiology, 88*(4), 664–672. http://dx.doi.org/10.1111/mmi.12203.

Beeby, M., Ribardo, D. A., Brennan, C. A., Ruby, E. G., Jensen, G. J., & Hendrixson, D. R. (2016). Diverse high-torque bacterial flagellar motors assemble wider stator rings using a conserved protein scaffold. *Proceedings of the National Academy of Sciences of the United States of America, 113*(13), E1917–E1926. http://dx.doi.org/10.1073/pnas.1518952113.

Biteen, J. S., Goley, E. D., Shapiro, L., & Moerner, W. E. (2012). Three-dimensional super-resolution imaging of the midplane protein FtsZ in live Caulobacter crescentus cells using astigmatism. *Chemphyschem, 13*(4), 1007–1012. http://dx.doi.org/10.1002/cphc.201100686.

Bonemann, G., Pietrosiuk, A., & Mogk, A. (2010). Tubules and donuts: A type VI secretion story. *Molecular Microbiology, 76*(4), 815–821. http://dx.doi.org/10.1111/j.1365-2958.2010.07171.x.

Briegel, A., Ames, P., Gumbart, J. C., Oikonomou, C. M., Parkinson, J. S., & Jensen, G. J. (2013). The mobility of two kinase domains in the Escherichia coli chemoreceptor array varies with signalling state. *Molecular Microbiology, 89*(5), 831–841. http://dx.doi.org/10.1111/mmi.12309.

Briegel, A., Beeby, M., Thanbichler, M., & Jensen, G. J. (2011). Activated chemoreceptor arrays remain intact and hexagonally packed. *Molecular Microbiology, 82*(3), 748–757. http://dx.doi.org/10.1111/j.1365-2958.2011.07854.x.

Briegel, A., Ladinsky, M. S., Oikonomou, C., Jones, C. W., Harris, M. J., Fowler, D. J., ... Jensen, G. J. (2014). Structure of bacterial cytoplasmic chemoreceptor arrays and implications for chemotactic signaling. *eLife, 3*, http://dx.doi.org/10.7554/eLife.02151.

Briegel, A., Li, X., Bilwes, A. M., Hughes, K. T., Jensen, G. J., & Crane, B. R. (2012). Bacterial chemoreceptor arrays are hexagonally packed trimers of receptor dimers networked by rings of kinase and coupling proteins. *Proceedings of the National Academy of Sciences of the United States of America, 109*(10), 3766–3771. http://dx.doi.org/10.1073/pnas.1115719109.

Briegel, A., Ortega, D. R., Tocheva, E. I., Wuichet, K., Li, Z., Chen, S., ... Jensen, G. J. (2009). Universal architecture of bacterial chemoreceptor arrays. *Proceedings of the National Academy of Sciences of the United States of America, 106*(40), 17181–17186. http://dx.doi.org/10.1073/pnas.0905181106.

Cassidy, C. K., Himes, B. A., Alvarez, F. J., Ma, J., Zhao, G., Perilla, J. R., ... Zhang, P. (2015). CryoEM and computer simulations reveal a novel kinase conformational switch in bacterial chemotaxis signaling. *eLife, 4*, e08419. http://dx.doi.org/10.7554/eLife.08419.

Chang, Y. W., Rettberg, L., Treuner-Lange, A., Iwasa, J., Sogaard-Andersen, L., & Jensen, G. J. (2016). Architecture of the type IVa pilus machine. *Science, 351*(6278), aad2001.

Chastanet, A., & Carballido-Lopez, R. (2012). The actin-like MreB proteins in Bacillus subtilis: A new turn. *Frontiers in Bioscience (Scholar Edition), 4*, 1582–1606.

Chen, S., Beeby, M., Murphy, G. E., Leadbetter, J. R., Hendrixson, D. R., Briegel, A., ... Jensen, G. J. (2011). Structural diversity of bacterial flagellar motors. *The EMBO Journal, 30*(14), 2972–2981. http://dx.doi.org/10.1038/emboj.2011.186.

Comolli, L. R., Kundmann, M., & Downing, K. H. (2006). Characterization of intact subcellular bodies in whole bacteria by cryo-electron tomography and spectroscopic imaging. *Journal of Microscopy, 223*(Pt 1), 40–52. http://dx.doi.org/10.1111/j.1365-2818.2006.01597.x.

Diestra, E., Fontana, J., Guichard, P., Marco, S., & Risco, C. (2009). Visualization of proteins in intact cells with a clonable tag for electron microscopy. *Journal of Structural Biology, 165*(3), 157–168. http://dx.doi.org/10.1016/j.jsb.2008.11.009.

Dmitriev, B. A., Toukach, F. V., Schaper, K. J., Holst, O., Rietschel, E. T., & Ehlers, S. (2003). Tertiary structure of bacterial murein: The scaffold model. *Journal of Bacteriology, 185*(11), 3458–3468.

Dominguez-Escobar, J., Chastanet, A., Crevenna, A. H., Fromion, V., Wedlich-Soldner, R., & Carballido-Lopez, R. (2011). Processive movement of MreB-associated cell wall biosynthetic complexes in bacteria. *Science, 333*(6039), 225–228. http://dx.doi.org/10.1126/science.1203466.

Dubochet, J., Adrian, M., Chang, J. J., Homo, J. C., Lepault, J., McDowall, A. W., & Schultz, P. (1988). Cryo-electron microscopy of vitrified specimens. *Quarterly Reviews of Biophysics, 21*(2), 129–228.

Dubochet, J., Zuber, B., Eltsov, M., Bouchet-Marquis, C., Al-Amoudi, A., & Livolant, F. (2007). How to "read" a vitreous section. *Methods in Cell Biology, 79*, 385–406. http://dx.doi.org/10.1016/S0091-679X(06)79015-X.

Farley, M. M., Hu, B., Margolin, W., & Liu, J. (2016). Minicells, back in fashion. *Journal of Bacteriology, 198*(8), 1186–1195. http://dx.doi.org/10.1128/JB.00901-15.

Fu, X., Himes, B. A., Ke, D., Rice, W. J., Ning, J., & Zhang, P. (2014). Controlled bacterial lysis for electron tomography of native cell membranes. *Structure, 22*(12), 1875–1882. http://dx.doi.org/10.1016/j.str.2014.09.017.

Gan, L., Chen, S., & Jensen, G. J. (2008). Molecular organization of Gram-negative peptidoglycan. *Proceedings of the National Academy of Sciences of the United States of America, 105*(48), 18953–18957. http://dx.doi.org/10.1073/pnas.0808035105.

Garner, E. C., Bernard, R., Wang, W., Zhuang, X., Rudner, D. Z., & Mitchison, T. (2011). Coupled, circumferential motions of the cell wall synthesis machinery and MreB filaments in B. subtilis. *Science, 333*(6039), 222–225. http://dx.doi.org/10.1126/science.1203285.

Goley, E. D., Yeh, Y. C., Hong, S. H., Fero, M. J., Abeliuk, E., McAdams, H. H., & Shapiro, L. (2011). Assembly of the Caulobacter cell division machine. *Molecular Microbiology, 80*(6), 1680–1698. http://dx.doi.org/10.1111/j.1365-2958.2011.07677.x.

Hayhurst, E. J., Kailas, L., Hobbs, J. K., & Foster, S. J. (2008). Cell wall peptidoglycan architecture in Bacillus subtilis. *Proceedings of the National Academy of Sciences of the United States of America, 105*(38), 14603–14608. http://dx.doi.org/10.1073/pnas.0804138105.

Henderson, G. P., & Jensen, G. J. (2006). Three-dimensional structure of Mycoplasma pneumoniae's attachment organelle and a model for its role in gliding motility. *Molecular Microbiology, 60*(2), 376–385. http://dx.doi.org/10.1111/j.1365-2958.2006.05113.x.

Holtje, J. V. (1998). Growth of the stress-bearing and shape-maintaining murein sacculus of Escherichia coli. *Microbiology and Molecular Biology Reviews*, *62*(1), 181–203.

Hu, B., Morado, D. R., Margolin, W., Rohde, J. R., Arizmendi, O., Picking, W. L., ... Liu, J. (2015). Visualization of the type III secretion sorting platform of Shigella flexneri. *Proceedings of the National Academy of Sciences of the United States of America*, *112*(4), 1047–1052. http://dx.doi.org/10.1073/pnas.1411610112.

Iancu, C. V., Ding, H. J., Morris, D. M., Dias, D. P., Gonzales, A. D., Martino, A., & Jensen, G. J. (2007). The structure of isolated Synechococcus strain WH8102 carboxysomes as revealed by electron cryotomography. *Journal of Molecular Biology*, *372*(3), 764–773. http://dx.doi.org/10.1016/j.jmb.2007.06.059.

Iancu, C. V., Morris, D. M., Dou, Z., Heinhorst, S., Cannon, G. C., & Jensen, G. J. (2010). Organization, structure, and assembly of alpha-carboxysomes determined by electron cryotomography of intact cells. *Journal of Molecular Biology*, *396*(1), 105–117. http://dx.doi.org/10.1016/j.jmb.2009.11.019.

Ingerson-Mahar, M., Briegel, A., Werner, J. N., Jensen, G. J., & Gitai, Z. (2010). The metabolic enzyme CTP synthase forms cytoskeletal filaments. *Nature Cell Biology*, *12*(8), 739–746. http://dx.doi.org/10.1038/ncb2087.

Jasnin, M., Asano, S., Gouin, E., Hegerl, R., Plitzko, J. M., Villa, E., ... Baumeister, W. (2013). Three-dimensional architecture of actin filaments in Listeria monocytogenes comet tails. *Proceedings of the National Academy of Sciences of the United States of America*, *110*(51), 20521–20526. http://dx.doi.org/10.1073/pnas.1320155110.

Jones, L. J., Carballido-Lopez, R., & Errington, J. (2001). Control of cell shape in bacteria: Helical, actin-like filaments in Bacillus subtilis. *Cell*, *104*(6), 913–922.

Khursigara, C. M., Lan, G., Neumann, S., Wu, X., Ravindran, S., Borgnia, M. J., ... Subramaniam, S. (2011). Lateral density of receptor arrays in the membrane plane influences sensitivity of the E. coli chemotaxis response. *The EMBO Journal*, *30*(9), 1719–1729. http://dx.doi.org/10.1038/emboj.2011.77.

Komeili, A., Li, Z., Newman, D. K., & Jensen, G. J. (2006). Magnetosomes are cell membrane invaginations organized by the actin-like protein MamK. *Science*, *311*(5758), 242–245. http://dx.doi.org/10.1126/science.1123231.

Konorty, M., Kahana, N., Linaroudis, A., Minsky, A., & Medalia, O. (2008). Structural analysis of photosynthetic membranes by cryo-electron tomography of intact Rhodopseudomonas viridis cells. *Journal of Structural Biology*, *161*(3), 393–400. http://dx.doi.org/10.1016/j.jsb.2007.09.014.

Koster, A. J., Grimm, R., Typke, D., Hegerl, R., Stoschek, A., Walz, J., & Baumeister, W. (1997). Perspectives of molecular and cellular electron tomography. *Journal of Structural Biology*, *120*(3), 276–308. http://dx.doi.org/10.1006/jsbi.1997.3933.

Kuhn, J., Briegel, A., Morschel, E., Kahnt, J., Leser, K., Wick, S., ... Thanbichler, M. (2010). Bactofilins, a ubiquitous class of cytoskeletal proteins mediating polar localization of a cell wall synthase in Caulobacter crescentus. *The EMBO Journal*, *29*(2), 327–339. http://dx.doi.org/10.1038/emboj.2009.358.

Kurner, J., Frangakis, A. S., & Baumeister, W. (2005). Cryo-electron tomography reveals the cytoskeletal structure of Spiroplasma melliferum. *Science*, *307*(5708), 436–438. http://dx.doi.org/10.1126/science.1104031.

Lan, G., Daniels, B. R., Dobrowsky, T. M., Wirtz, D., & Sun, S. X. (2009). Condensation of FtsZ filaments can drive bacterial cell division. *Proceedings of the National Academy of Sciences of the United States of America*, *106*(1), 121–126. http://dx.doi.org/10.1073/pnas.0807963106.

Landgraf, D., Okumus, B., Chien, P., Baker, T. A., & Paulsson, J. (2012). Segregation of molecules at cell division reveals native protein localization. *Nature Methods*, *9*(5), 480–482. http://dx.doi.org/10.1038/nmeth.1955.

Li, Z., Trimble, M. J., Brun, Y. V., & Jensen, G. J. (2007). The structure of FtsZ filaments in vivo suggests a force-generating role in cell division. *The EMBO Journal*, *26*(22), 4694–4708. http://dx.doi.org/10.1038/sj.emboj.7601895.

Liu, J., Hu, B., Morado, D. R., Jani, S., Manson, M. D., & Margolin, W. (2012). Molecular architecture of chemoreceptor arrays revealed by cryoelectron tomography of Escherichia coli minicells. *Proceedings of the National Academy of Sciences of the United States of America*, *109*(23), E1481–E1488. http://dx.doi.org/10.1073/pnas.1200781109.

Liu, J., Lin, T., Botkin, D. J., McCrum, E., Winkler, H., & Norris, S. J. (2009). Intact flagellar motor of Borrelia burgdorferi revealed by cryo-electron tomography: Evidence for stator ring curvature and rotor/C-ring assembly flexion. *Journal of Bacteriology*, *191*(16), 5026–5036. http://dx.doi.org/10.1128/JB.00340-09.

Liu, J., McBride, M. J., & Subramaniam, S. (2007). Cell surface filaments of the gliding bacterium Flavobacterium johnsoniae revealed by cryo-electron tomography. *Journal of Bacteriology*, *189*(20), 7503–7506. http://dx.doi.org/10.1128/JB.00957-07.

Lu, C., Reedy, M., & Erickson, H. P. (2000). Straight and curved conformations of FtsZ are regulated by GTP hydrolysis. *Journal of Bacteriology*, *182*(1), 164–170.

Ma, X., Ehrhardt, D. W., & Margolin, W. (1996). Colocalization of cell division proteins FtsZ and FtsA to cytoskeletal structures in living Escherichia coli cells by using green fluorescent protein. *Proceedings of the National Academy of Sciences of the United States of America*, *93*(23), 12998–13003.

Marko, M., Hsieh, C., Schalek, R., Frank, J., & Mannella, C. (2007). Focused-ion-beam thinning of frozen-hydrated biological specimens for cryo-electron microscopy. *Nature Methods*, *4*(3), 215–217. http://dx.doi.org/10.1038/nmeth1014.

Mercogliano, C. P., & DeRosier, D. J. (2007). Concatenated metallothionein as a clonable gold label for electron microscopy. *Journal of Structural Biology*, *160*(1), 70–82. http://dx.doi.org/10.1016/j.jsb.2007.06.010.

Murphy, G. E., Leadbetter, J. R., & Jensen, G. J. (2006). In situ structure of the complete Treponema primitia flagellar motor. *Nature*, *442*(7106), 1062–1064. http://dx.doi.org/10.1038/nature05015.

Palade, G. E. (1952). A study of fixation for electron microscopy. *The Journal of Experimental Medicine*, *95*(3), 285–298.

Park, S. Y., Borbat, P. P., Gonzalez-Bonet, G., Bhatnagar, J., Pollard, A. M., Freed, J. H., ... Crane, B. R. (2006). Reconstruction of the chemotaxis receptor-kinase assembly. *Nature Structural & Molecular Biology*, *13*(5), 400–407. http://dx.doi.org/10.1038/nsmb1085.

Pilhofer, M., Aistleitner, K., Ladinsky, M. S., Konig, L., Horn, M., & Jensen, G. J. (2014). Architecture and host interface of environmental chlamydiae revealed by electron cryotomography. *Environmental Microbiology*, *16*(2), 417–429. http://dx.doi.org/10.1111/1462-2920.12299.

Pilhofer, M., & Jensen, G. J. (2013). The bacterial cytoskeleton: More than twisted filaments. *Current Opinion in Cell Biology*, *25*(1), 125–133. http://dx.doi.org/10.1016/j.ceb.2012.10.019.

Pilhofer, M., Ladinsky, M. S., McDowall, A. W., & Jensen, G. J. (2010). Bacterial TEM: New insights from cryo-microscopy. *Methods in Cell Biology*, *96*, 21–45. http://dx.doi.org/10.1016/S0091-679X(10)96002-0.

Psencik, J., Collins, A. M., Liljeroos, L., Torkkeli, M., Laurinmaki, P., Ansink, H. M., ... Butcher, S. J. (2009). Structure of chlorosomes from the green filamentous bacterium Chloroflexus aurantiacus. *Journal of Bacteriology*, *191*(21), 6701–6708. http://dx.doi.org/10.1128/JB.00690-09.

Pukatzki, S., Ma, A. T., Sturtevant, D., Krastins, B., Sarracino, D., Nelson, W. C., ... Mekalanos, J. J. (2006). Identification of a conserved bacterial protein secretion system in Vibrio cholerae using the Dictyostelium host model system. *Proceedings of the National Academy of Sciences of the United States of America*, *103*(5), 1528–1533. http://dx.doi.org/10.1073/pnas.0510322103.

Salje, J., van den Ent, F., de Boer, P., & Lowe, J. (2011). Direct membrane binding by bacterial actin MreB. *Molecular Cell*, *43*(3), 478–487. http://dx.doi.org/10.1016/j.molcel.2011.07.008.

Salje, J., Zuber, B., & Lowe, J. (2009). Electron cryomicroscopy of E. coli reveals filament bundles involved in plasmid DNA segregation. *Science*, *323*(5913), 509–512. http://dx.doi.org/10.1126/science.1164346.

Scheffel, A., Gruska, M., Faivre, D., Linaroudis, A., Plitzko, J. M., & Schuler, D. (2006). An acidic protein aligns magnetosomes along a filamentous structure in magnetotactic bacteria. *Nature*, *440*(7080), 110–114. http://dx.doi.org/10.1038/nature04382.

Schlimpert, S., Klein, E. A., Briegel, A., Hughes, V., Kahnt, J., Bolte, K., ... Thanbichler, M. (2012). General protein diffusion barriers create compartments within bacterial cells. *Cell*, *151*(6), 1270–1282. http://dx.doi.org/10.1016/j.cell.2012.10.046.

Schmid, M. F., Paredes, A. M., Khant, H. A., Soyer, F., Aldrich, H. C., Chiu, W., & Shively, J. M. (2006). Structure of Halothiobacillus neapolitanus carboxysomes by cryo-electron tomography. *Journal of Molecular Biology*, *364*(3), 526–535. http://dx.doi.org/10.1016/j.jmb.2006.09.024.

Seybert, A., Herrmann, R., & Frangakis, A. S. (2006). Structural analysis of Mycoplasma pneumoniae by cryo-electron tomography. *Journal of Structural Biology*, *156*(2), 342–354. http://dx.doi.org/10.1016/j.jsb.2006.04.010.

Shih, Y. L., Le, T., & Rothfield, L. (2003). Division site selection in Escherichia coli involves dynamic redistribution of Min proteins within coiled structures that extend between the two cell poles. *Proceedings of the National Academy of Sciences of the United States of America*, *100*(13), 7865–7870. http://dx.doi.org/10.1073/pnas.1232225100.

Shimizu, T. S., Le Novere, N., Levin, M. D., Beavil, A. J., Sutton, B. J., & Bray, D. (2000). Molecular model of a lattice of signalling proteins involved in bacterial chemotaxis. *Nature Cell Biology*, *2*(11), 792–796. http://dx.doi.org/10.1038/35041030.

Sobhanifar, S., King, D. T., & Strynadka, N. C. (2013). Fortifying the wall: Synthesis, regulation and degradation of bacterial peptidoglycan. *Current Opinion in Structural Biology*, *23*(5), 695–703. http://dx.doi.org/10.1016/j.sbi.2013.07.008.

Swulius, M. T., Chen, S., Jane Ding, H., Li, Z., Briegel, A., Pilhofer, M., ... Jensen, G. J. (2011). Long helical filaments are not seen encircling cells in electron cryotomograms of rod-shaped bacteria. *Biochemical and Biophysical Research Communications*, *407*(4), 650–655. http://dx.doi.org/10.1016/j.bbrc.2011.03.062.

Swulius, M. T., & Jensen, G. J. (2012). The helical MreB cytoskeleton in Escherichia coli MC1000/pLE7 is an artifact of the N-terminal yellow fluorescent protein tag. *Journal of Bacteriology*, *194*(23), 6382–6386. http://dx.doi.org/10.1128/JB.00505-12.

Szwedziak, P., Wang, Q., Bharat, T. A., Tsim, M., & Lowe, J. (2014). Architecture of the ring formed by the tubulin homologue FtsZ in bacterial cell division. *eLife*, *4*, e04601. http://dx.doi.org/10.7554/eLife.04601.

References

Ting, C. S., Hsieh, C., Sundararaman, S., Mannella, C., & Marko, M. (2007). Cryo-electron tomography reveals the comparative three-dimensional architecture of Prochlorococcus, a globally important marine cyanobacterium. *Journal of Bacteriology*, *189*(12), 4485–4493. http://dx.doi.org/10.1128/JB.01948-06.

Tocheva, E. I., Dekas, A. E., McGlynn, S. E., Morris, D., Orphan, V. J., & Jensen, G. J. (2013). Polyphosphate storage during sporulation in the gram-negative bacterium Acetonema longum. *Journal of Bacteriology*, *195*(17), 3940–3946. http://dx.doi.org/10.1128/JB.00712-13.

Tocheva, E. I., Lopez-Garrido, J., Hughes, H. V., Fredlund, J., Kuru, E., Vannieuwenhze, M. S., ... Jensen, G. J. (2013). Peptidoglycan transformations during Bacillus subtilis sporulation. *Molecular Microbiology*, *88*(4), 673–686. http://dx.doi.org/10.1111/mmi.12201.

van Teeffelen, S., Wang, S., Furchtgott, L., Huang, K. C., Wingreen, N. S., Shaevitz, J. W., & Gitai, Z. (2011). The bacterial actin MreB rotates, and rotation depends on cell-wall assembly. *Proceedings of the National Academy of Sciences of the United States of America*, *108*(38), 15822–15827. http://dx.doi.org/10.1073/pnas.1108999108.

Wang, Q., Mercogliano, C. P., & Lowe, J. (2011). A ferritin-based label for cellular electron cryotomography. *Structure*, *19*(2), 147–154. http://dx.doi.org/10.1016/j.str.2010.12.002.

Werner, J. N., Chen, E. Y., Guberman, J. M., Zippilli, A. R., Irgon, J. J., & Gitai, Z. (2009). Quantitative genome-scale analysis of protein localization in an asymmetric bacterium. *Proceedings of the National Academy of Sciences of the United States of America*, *106*(19), 7858–7863. http://dx.doi.org/10.1073/pnas.0901781106.

Zemlin, F., Weiss, K., Schiske, P., Kunath, W., & Hermann, K.-H. (1978). Coma-free alignment of high resolution electron microscopes with the aid of optical diffractograms. *Ultramicroscopy*, *3*, 49–60.

Index

Note: Page numbers followed by "*f*" indicate figures, and "*t*" indicate tables.

A

Agarose, 93–94, 93*f*, 100
   pads with culture media, 69
   and PDMS, 77, 92–94
Atomic force microscopy (AFM), 132–133
   in bacterial cell wall, 16–17
   peptidoglycan in *E. coli*, 16–17, 16–17*f*
AutoCAD software, 72
Autolysins, 7

B

Bacillus subtilis, 94–96
   AFM, 132–133
   dacA deletion strains, 34
   ECT, 132*f*
   fluorescent ramoplanin, 23
   PG architecture, 17, 18*f*
   phase contrast images, 19*f*
   sporulation, 120–121
   Van-FL labelling in, 21–22
Bacterial cell biology, 116–118
Bacterial cell wall, 16–17, 94–96
Bacterial chemoreceptor array,
     architecture, 124*f*
Bacterial microfluidics systems, 96–97
   application, 69, 71*t*
   biofilm, 86–90, 88*f*
   cell cycle analysis, 90–94
   cell shape, 94–96
   computational analysis and control of, 80–83, 81–82*t*
   designing and setting up, 79–80
   droplet-based systems, 83
   engineering, 83–86
   fabrication, 71–77
   fluid flow, 77–79, 79*f*
   geometry study, 94–96
   methods, 97–105
   microbial ecology, 86–90
   size homeostasis studies, 90–94
   spatial arrangements and interactions, 90
   species-species interactions, 86–90
Biofilm, 86–90, 88*f*
BioFlux system, 87–89, 88*f*
Bio-MEMS, 69–70

C

Caulobacter crescentus, 14–16, 116–118, 122–125, 129
Cell biology of bacteria, 116–118
Cell culture, 56–57
Cell cycle, 90–94, 129–130
Cell lineage, time-lapse microscopy, 61–63, 61*f*
   cell tracking, 61–62
   error correction, 63
   error detection, 62–63
   experimental results, 64–66
   Recall–Precision curve, 64, 65*f*
   voting mechanism, 63, 64*f*, 66
Cell shape, 94–96
Cell size control, 90
Cell tracking, 61–62
Cell wall, 13–14, 16–17
CET. *See* Cryoelectron tomography (CET)
Chlamydia trachomatis, 38–40, 41*f*
COMSOL, 74, 75*f*
Correlated light and electron microscopy (CLEM), 125–127
Cryoelectron tomography (CET), 14–16, 15*f*
Cryogenic electron microscopy (Cryo-EM), 13, 116, 130
Cryo-TEM, *E. coli*, 13–14, 14*f*

D

D-amino acid-based fluorescent probes, 25–27, 25*f*
Differential interference contrast (DIC) microscopy, 19

E

E-beam lithography, 74–76
Eigen image, 59
Electron cryotomography (ECT), 115–116, 117*f*
   advantages, 129–133
   bacterial chemoreceptor array, 124*f*
   bacterial nanomachine, 119–120*f*
   FtsZ filaments structure, 126–127*f*
   gram-positive cell wall, 132*f*
   identification, 125–127
   intact cells, 130
   limitations, 116, 120–128
   microbiology, 116–120
   missing wedge, 122–125

141

Electron cryotomography (ECT) (*Continued*)
  MreB structure, 123f
  native proteins, 129–130
  radiation damage, 121–122
  static snapshots, 127–128
  thickness, 120–121
  3D, 131–133
Electron microscopy (EM), 115
  peptidoglycan in *E. coli*
    AFM, 16–17, 16–17f
    cryoelectron tomography, 14–16, 15f
    cryo-TEM, 13–14, 14f
    elucidation, 11–13
    revealing bacterial morphology, 9–10
  Spirillum serpens, 11f
  streptococcal cells, 10f
Escherichia coli, 85–86
  cell shape, 94–96
  ECT, 122–125, 129
  image analysis, 57–66
  immunoelectron microscopy, 12–13, 12f
  Ldts in, 34t
  PBPs in, 32t
  peptidoglycan, 5f, 9f, 11–13, 24
    AFM, 16–17, 16–17f
    cryoelectron tomography, 14–16, 15f
    cryo-TEM, 13–14, 14f
    elucidation, 11–13
    FDAAs, 24–34, 25f, 29f
    fluorescent antibiotics, 21–23, 22–23f
    FWGA, 23–24
    operation and application, 18–21
    revealing bacterial morphology, 9–10
  septal region, 34–36
  time-lapse microscopy, 49–57

F
Fabrication of microfluidic devices, 71–72
  design, 72–74
  microfluidics chip, 77
  silicon wafers, 74–77
Fluid flow, bacterial microfluidics systems, 77–79, 79f
Fluorescence microscopy (FM), 20–21
Fluorescent antibiotics, 21–23, 22–23f
Fluorescent D-amino acids (FDAAs)
  applications, 34–40
  D-amino acid-based fluorescent probes, 25–27, 25f
  dipeptide derivatives, 40f
  modular syntheses, 29f
  optical microscopy, 24–34, 25f, 29f
  PG-specific labelling, 24–40, 26f
  potential pathways, 30f
  proposed mechanism of labelling, 27–34
  Streptococcus pneumoniae, 36–37, 37f
Fluorescent image, 60–61, 60f, 65
Fluorescent probes, D-amino acid-based, 25–27, 25f
Fluorescent wheat germ agglutinin (FWGA), 23–24

I
Image preprocessing, 57–58
Immunoelectron microscopy, 12–13, 12f
Intact cells, ECT, 130

L
Light microscopy, 115
Lipid II synthesis, 6

M
MicrobeTracker, 80
Microbial ecology, 86–90
Microbiology, ECT, 116–120
Microelectromechanical systems (MEMS), 69–70
Microfluidics
  for bacterial imaging (*see* Bacterial microfluidics systems)
  chip, 72, 73f, 77–78
  field of, 70
Miniaturised total analysis systems (μTASs), 69–70
Missing wedge, ECT, 122–125
Mother machine, 50
  advantages, 50–51
  cell culture and loading, 56–57
  dead-end channels, 50–53
  design considerations, 51–53
  general principles, 50–51, 51f
  LED-based systems, 54
  microfluidic setup, 57
  PDMS and, 55, 91–92, 92f, 97
  phase-contrast image, 53f
  setting up fluidic system, 55–56
  time-lapse microscopy, 53–54, 57
Myxococcus xanthus, 118, 119–120f

N
Native proteins, ECT, 129–130
Numerical aperture (NA), 19–20

O
Optical microscopy, peptidoglycan
  FDAAs, 24–34, 25f, 29f
  fluorescent antibiotics, 21–23, 22–23f
  FWGA, 23–24
  operation and application, 18–21

P

Park's nucleotide, 6
PDMS. *See* Polydimethylsiloxane (PDMS)
Penicillin-binding proteins (PBPs), 6–7
   in *E. coli*, 32*t*, 33*f*
   high-molecular weight, 29–30
   low-molecular weight, 29–30
   transpeptidation mechanisms, 31*f*
Peptidoglycan (PG), 3–4
   amino acid variations in, 7, 8*t*
   biosynthesis, 6–7, 9*f*
   B. subtilis, 17, 18*f*
   Chlamydia trachomatis, 38–40, 41*f*
   E. coli, 5*f*, 9*f*, 11–13, 24
      electron microscopy, 9–17
      optical microscopy, 18–34
   insertion and remodelling, 6–7
   lipid II synthesis, 6
   Park's nucleotide, 6
   structural variations in, 7–9
   structure and configuration, 4–5
Peristaltic pumps, 77–78
Phase-contrast image, 19*f*, 53*f*, 59, 59*f*
Phase contrast microscopy (PCM), 18–19
Point spread function (PSF), 19–20
Polydimethylsiloxane (PDMS)
   agarose and, 77, 92–94
   microfluidics chips, 72, 76, 79
      cleaning, 78
      production, 77
      soft lithography method, 79
      stages, 73*f*
   mother machine, 55, 91–92, 92*f*, 97
   single-cell chemostat, 93–94, 93*f*

R

Radiation damage, ECT, 121–122
Random Forest, 62–64, 66
Recall–Precision curve, 64, 65*f*
Reynolds number, 74

S

Scotch tape, 57
Silicon wafers, 74–77
Single-cell chemostat, 93–94, 93*f*, 100
Soft lithography method, 55, 77, 79
Species-species interactions, 86–90
SpotFinder tool, 80
Stanford microfluidics, 72
Staphylococcus aureus, 7
Streptococcus pneumoniae, PG
   fluorescent molecular probes, 22
   recycling/remodelling, 36–37
Super-resolution light microscopy, 80, 127, 130
Synthetic biology, 83–84
Syringe pumps, 77–78

T

Tanner L-edit software, 72
Time-lapse microscopy, 49–50
   cell lineage approaches, 61–63, 61*f*
      cell tracking, 61–62
      error correction, 63
      error detection, 62–63
      experimental results, 64–66
      Recall–Precision curve, 64, 65*f*
      voting mechanism, 63, 64*f*, 66
   fluorescent image, 60–61, 60*f*
   image analysis, 66
   image preprocessing, 57–58, 58*f*
   mother machine, 53–54, 57
   phase-contrast image, 59, 59*f*
   segmentation approaches, 58–61
   single-cell, 36–37, 38*f*, 57–66
Transmission electron microscope (TEM), 116–118

V

Vibrio cholerae, 125–127, 128*f*